The Silent Epidemic

The Silent Epidemic
Coal and the Hidden Threat to Health

Alan H. Lockwood, MD

The MIT Press
Cambridge, Massachusetts
London, England

MIT Press books may be purchased at special quantity discounts for business or sales promotional use. For information, please email special_sales@mitpress.mit.edu or write to Special Sales Department, The MIT Press, 55 Hayward Street, Cambridge, MA 02142.

This book was set in Sabon by Toppan Best-set Premedia Limited. Printed and bound in the United States of America.

Library of Congress Cataloging-in-Publication Data

Lockwood, Alan H.
The silent epidemic : coal and the hidden threat to health / Alan H. Lockwood.
  p. cm.
Includes bibliographical references and index.
ISBN 978-0-262-01789-3 (hardcover : alk. paper)
1. Coal–Environmental aspects. 2. Pollution–Health aspects. I. Title.
TD195.C58L63 2012
363.17′9–dc23
2011053211

10 9 8 7 6 5 4 3 2 1

For Owen, Clara, and Duncan
from whom we have borrowed this earth

Contents

Preface and Acknowledgments

The Silent Epidemic: Coal and the Hidden Threat to Health is the outgrowth of the white paper *Coal's Assault on Human Health* published by Physicians for Social Responsibility (PSR) in the fall of 2009. At the completion of that project, it was evident that there was much more to be told. All of us who worked on this project were astounded by the number and variety of diseases linked to the pollutants produced by burning coal and the enormous number of people around the world whose lives are diminished or curtailed by these pollutants. As a measure of respect and admiration for this organization, the royalties from this book will be donated to Physicians for Social Responsibility so that it can continue its important work.

Although I was the principal author of *Coal's Assault on Human Health*, I had very able help from Molly Rauch, a gifted writer and editor as well as Kristin Welker-Hood and Barbara Gottlieb. All three of these dedicated professionals were on the staff of Physicians for Social Responsibility. Sally Murray James was our able graphic artist. That project was made possible by a generous grant from The Energy Foundation. I am indebted to them for their early support that led to this project.

I am enormously grateful to the extraordinary efforts of so many gifted scientists and physicians upon whose work I relied. To the authors whose work I have not cited, I offer my apology and thanks. Even book authors are limited. In many instances I relied on pertinent reviews or seminal papers. These often become the source for much of the information in the remainder of the paragraph. Almost every section of every chapter could be supported by near-astronomical numbers of peer-reviewed papers published in leading medical journals and other reports.

I wish to extend a special thank you to the American Academy of Neurology. As the program director of the adult neurology residency program at the University at Buffalo, I was eligible for training in the principles of evidence-based medicine. I drew heavily on this valuable experience, particularly in chapter 7, on basic health considerations, and the ensuing chapters dealing with diseases of the respiratory, cardiovascular, and nervous systems, and the chapter on emerging health considerations. Amy Wallace and Gary Gronseth, MD were particularly helpful.

Physicians for Social Responsibility's mission is "to protect human life from the gravest threats to health and survival." Burning coal poses that threat. No one who has been downwind of burning coal would doubt that inhaling this smoke is not good for you. John Evelyn knew that in 1661 when he wrote his treatise, "Fumifugium," one of the earliest works on air pollution. In this book I have kept much of the fundamental organization and focus of *Coal's Assault on Human Health*. Physicians are accustomed and trained to think in terms of organ systems. This is the rationale for the chapters devoted to diseases of the lungs, heart and vascular system, the nervous system, and others. It is increasingly evident that the coal-pollutant story is incomplete. Recent studies link diseases such as type 2 diabetes mellitus and Alzheimer's disease to the pollutants produced by burning coal. It seems almost certain that as epidemiological and other research progresses, even more disease states will be linked to burning coal.

To the extent that this was possible I have relied on papers published in the peer-reviewed scientific literature during the preparation of this book. I have intentionally avoided the good work product of the many nonprofits whose efforts have made this earth a healthier place. This was not because of a lack of respect for their efforts, but rather to avoid perceptions of bias. In many instances I have relied on reports prepared by governmental agencies. Frequently they are the only source of important data. Many of these reports are peer-reviewed.

Occasionally there is reluctance to quote items of scholarship that are not peer-reviewed. This book has been subjected to more reviews by my colleagues than almost anything else I have written. The process began with an informal contact with Clay Morgan, the Senior Acquisition Editor for Environmental Sciences at the MIT Press. He provided early encouragement and belief in the value of the book. His cogent advice helped steer me through the remaining steps leading to eventual publication. The first of these was a detailed table of contents, description of the target audience, a curriculum vitae, and other information related to the proposed book. The PSR report was acceptable in lieu of sample chapters. This information was submitted to the first review by peers. My responses to the reviews and the other items described above were then submitted to a publication committee. Only then was I offered a contract. A second peer review was performed after the book draft was finished. The revision took these reviews into account along with final editing done by Dana Andrus at the MIT Press. The entire process, start to finish, took slightly more than a year and a half.

In writing this book, I was charged with the task of making it accessible to as many readers as possible, ranging from physicians and scientists, to politicians, policy makers, and last, but clearly not least, to those who care about health and the environment. I have avoided most of the acronyms that make so many publications confusing to too many readers. Some of my readers will, rightfully, demand indications of the statistical validity of results I present.

For those who are not accustomed to statistical inferences and their meaning, I have tried to deal with the presentation of results to the best of my ability. I have placed some of the statistical data, such as confidence intervals, in parentheses, realizing that this interrupts the flow of the text for some but are welcome data for others. Hopefully the glossary and some of the explanations provided in chapter 7 will be helpful to readers.

I am especially grateful to my wife, Anne Lockwood, for her multiple readings and editorial suggestions and to Michael Cohen and Patricia Pliner who gave generously of their time to critique an early draft of the book. They were extremely helpful in the task of helping me to see the forest amid the trees. However, the responsibility for the final content is squarely on my shoulders.

Finally, the task of writing a book is formidable. I am extraordinarily fortunate to have the love and support of my wife Anne. Without her, this book would not exist.

Alan H. Lockwood
Buffalo, NY

1

Introduction

Linking Coal and Health

We are in the midst of a silent epidemic caused by the exposure to coal-derived pollutants. I refer to this as a silent epidemic because too few of us are aware of the relationship between coal and health, the focus of this book. Each year this epidemic claims the lives of tens of thousands of Americans and causes hundreds of thousands of serious and minor illnesses. You will not find "exposure to coal-derived pollution" on a single death certificate. The same is true for smoking cigarettes, yet the links to both are there, but much less visible and well known for coal. Instead of blaming coal, vital statistics chronicle deaths due to asthma, chronic obstructive pulmonary disease, lung and other cancers, heart attacks and other cardiovascular diseases such as stroke, sudden cardiac death due to heart rhythm disturbances, and an ever-lengthening list of other major causes of death and disease. The list will grow as the effects of global warming become more pronounced.

In 1971 President Richard Nixon signed the National Cancer Act, marking the beginning of what is now referred to as the "War on Cancer" the second ranked cause of death among Americans in 2008. We have not followed suit with regard to pollution. Far from it, in recent years it is the other way around. Those who seek to preserve and extend the health benefits that follow the control of pollution are under assault by those who deny the validity of the peer-reviewed, science-based evidence that pollution damages health. They have focused their attacks on the Environmental Protection Agency (EPA), the branch of the government specifically charged with protecting human health and the environment. This antagonism has become a litmus test for conservative Republicans who are egged on by members of the Tea Party. Ironically the EPA was founded and the critical Clean Air Act Amendments of 1990 were adopted during Republican administrations.

The evidence linking coal and disease is strong and getting stronger each year. Recent peer-reviewed articles in major medical journals provide surprising estimates of the morbidity and mortality associated with burning coal. European data reported in the prestigious medical journal, *The Lancet*, show that 24.5 deaths are expected for each TerraWatthour (TWh = 10^{12} Watthours) of electricity generated (95% CI = 6.1–98), in addition to 225 serious illnesses (95% CI = 56.2–899), and 13,288 minor illnesses (95% CI = 3,322–53,150) [1]. Burning lignite, a lower rank or type of coal that yields more pollutants than bituminous coal, raises these numbers to 32.6 deaths (95% CI = 8.2–130), 298 serious illnesses (95% CI = 74.6–1,193), and 17,676 minor illnesses (95% CI = 4,419–70,704) for each TWh. To give these data perspective, consider the fact that nearly half of the electricity generated in the United States in 2007 came from coal-fired power plants. If these European estimates are applied to the United States, as many as 50,000 deaths per year may be attributable to burning coal. Although differences in the population density between Europe and the United States are substantial, and there are large boundaries on the 95% confidence limits associated with these data, it is clear that burning coal has major adverse health effects that must not be ignored.

As a scientist, I have a certain disdain for arguments that begin with the phrase, "it is intuitively obvious that . . ." However, for coal, a certain amount of intuition is warranted. Burning coal imparts a terrible smell to the air. It can't be good for you. Almost anyone who has been downwind of burning coal knows this. Presumably, this is what led John Evelyn to warn his "Sacred Majestie" the king, of the "hazzard to Your Health" and that "kindled this indignation of [his] against it" [2]. This 1661 treatise may be the earliest reference to the adverse health effects associated with burning coal.

Systematic studies of the relationship between burning coal, industrial activity and the health effects of hazardous air pollutants date clearly to 1872 with the publication of *Air and Rain: The Beginning of Chemical Climatology* by Robert Angus Smith (cited by [3]). Since then there have been a number of sentinel events that link episodes of severe air pollution to a variety of illnesses [3]. In October 1948, almost half of the 14,000 residents of Donora, Pennsylvania, were sickened when atmospheric conditions trapped toxic emissions from a nearby smelter: 20 died and 400 required hospitalizations. In 1952, the infamous killer fog in London sent death rates and hospital admissions soaring. Overall hospital admissions increased by 43%. Those due to respiratory diseases rose by 163%. Almost 12,000 deaths were attributed to this environmental disaster caused, in part, by burning coal. The link between burning coal and adverse health was made strikingly clear in Dublin, Ireland, in the 1990s [4]. Because of increases in the cost of fuel oil in the 1980s, Dubliners switched from oil to bituminous coal to heat their homes and provide hot water. Subsequent increases in air pollution were associated with an increase in in-hospital deaths due to respiratory diseases. This led the Irish government to ban the "marketing, sale, and distribution of bituminous coal" on September 1, 1990. In the year that followed, black smoke concentrations declined by 70%, respiratory deaths fell by 15.5%, and cardiovascular deaths fell by 10.3% [4]. Approximately 450 lives were saved that year by this measure and hundreds of acute illnesses were prevented. Although burning coal was not the only cause of these epidemics, it was clearly a major factor in the production of the complex mixture of airborne pollutants that had diverse adverse effects on human health.

Mining, transporting, burning, and disposing of the products of the combustion of coal, all have major impacts on our health. With the passage of time, more and more adverse health effects have been attributed to our reliance on coal. Based on recent trends, it is likely that as time passes more and more adverse health effects will be shown to be the result of coal-derived pollutants.

Coal fueled the industrial revolution and has become an essential source of energy in virtually every country. Coal is the principal source of energy we use in the United States to generate electricity. About 45% of the energy used to generate electricity in the United States comes from burning coal, as shown in figure 1.1. In 2009 Americans paid an average of 9.82 cents per kWh to purchase 3,597 TWh of electricity [5]. These facts make coal a major component of the economy and form the nidus around which political, economic, health, and environmental considerations coalesce.

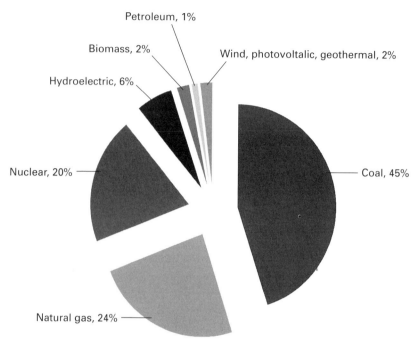

Figure 1.1
Sources of energy used for generation of electricity in 2010. Source US Energy Information Administration [9].

At the end of 2009, worldwide demonstrated coal reserves were approximately 908 billion tons [6]. Of that amount 28.9% are in the United States, 19% are in the Russian Federation, and 13.9% are in China. In 2006, the US electric power industry burned 1.026 billion tons of coal. Nearly 10% of this total was burned in Texas, with Missouri, Illinois, Indiana, Ohio, and Pennsylvania following closely behind (see figure 1.2). Increases in the cost of energy reinforce the 2006 Department of Energy estimate that 153 additional coal-fired power plants will be built by 2025. Powerful, well-funded lobbyists support plans to build approximately 100 of these units adding to the approximately 600 large coal-burning power plants already in existence.

Many of the pollutants produced by burning coal were listed by the US Environmental Protection Agency in its 1998 Report to Congress [7]. This report identified as many as 67 different hazardous air pollutants. This list did not include three pollutants, particulate matter and oxides of nitrogen

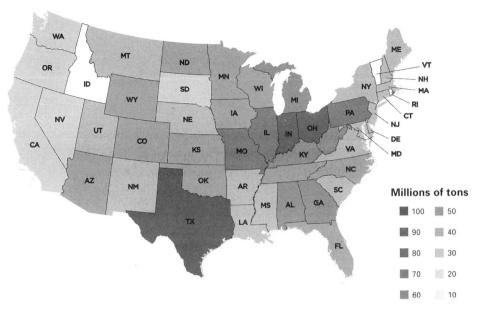

Figure 1.2
Coal consumption by coal-fired power plants, state by state. Texas led the nation with other states in the Ohio Valley that are close to coal fields and water transportation, following closely behind. Source: US Energy Information Association, 2009 data [10].

and sulfur, that along with carbon monoxide, lead, and ozone are defined by the EPA as criteria pollutants because of their threats to human health and the environment. The Clean Air Act requires the EPA to establish National Ambient Air Quality Standards for these six criteria pollutants because of their impacts on health and the environment and empowers the Agency to make rules that are necessary to meet these standards. A discussion of these threats to health posed by coal forms the basis for this book.

Clean Coal

Because of the importance of energy in our economy and the large role that coal plays in the production of energy, politics, huge sums of money, information, and disinformation abound. "Clean coal" is at the center of many of these discussions. In its original usage, this term was used to refer to technologies that were designed to reduce emission of pollutants associated with burning coal, such as washing coal at the mine. This step removes some of the sulfur and other contaminants, including rocks and soil. This makes coal cleaner and cheaper to transport. More recently the definition of clean coal has been expanded to include carbon capture and storage.

The clean coal initiative gained status toward the 1980s and in early 1990s through the US Department of Energy's Clean Coal Technology and The Clean Coal Power Initiative. Information about these programs is available on line at http://www.fossil.energy.gov/programs/powersystems/cleancoal. Initially the initiative focused on reducing emissions of sulfur and nitrogen dioxides and their contribution to acid rain because of the effects they had on forests and waterways in the United States. The program has evolved with time and the realization that mercury and fine particles had serious adverse health effects and that carbon dioxide emissions were causing global warming.

The Department of Energy website touts the success of the program, since "more than 20 of the technologies tested in the original program achieved commercial success." However, success was not uniform. Among the eight first-round projects funded in 2002, only three were completed. The others were withdrawn or discontinued. Among the four second-round projects, three are active and one has been withdrawn.

Examples of projects include retrofitting a Wisconsin plant with the TOXECON™ system to control mercury, particulate, and oxides of sulfur and nitrogen; installing a waste-heat drier to remove moisture from lignite, a low-ranked coal, at a North Dakota plant to improve combustion efficiency; and employing advanced computational strategies to improve peak performance, remove soot, and control emissions at a Boston, Massachusetts, utility.

Clean coal got a shot in the arm from President Barak Obama whose February 3, 2010, memorandum to 14 federal agencies included the following statement: "Rapid commercial development and deployment of clean coal technologies, particularly carbon capture and storage (CCS), will help position the United States as a leader in the global clean energy race." There is an extensive discussion of CCS in chapter 5

In his book, *Big Coal,* Jeff Goodell paints a picture of the influence of coal on the political system in the United States, primarily through contributions to Republicans [8]. He draws on reports by the Center for

Responsive Politics when he concludes that coal-based companies spend large amounts on political contributions. For example, he cites data for Peabody Energy, a mining company that spends 0.65% of its profits on political contributions, a fraction greater than spent on the defense contractor Martin Marietta, Exxon/Mobil, Pfizer, and General Motors. Republicans received about ten times more money than Democrats when Peabody Energy opened its wallet. Goodell describes the Southern Company, a utility, as a "ruthless capitalist with a happy face" as he details contributions to the arts, education, and other worthy causes in an effort to, as he wrote, "disarm critics."

Coal companies have funded other campaigns designed to build support for their industry and deflect challenges to the pollution from coal. Television commercials and billboards are designed specifically for this purpose, some of which can be seen on YouTube. Spoofs are present too. One of these videos is titled "Get Clean Coal Clean" in which a housewife sprays a gray mist throughout her home while the narrator tells us that "clean coal harnesses the awesome power of the word clean."

Make no mistake. Billions of dollars and tens of thousands of lives each year depend on the outcome of clean coal campaigns.

Advice to Readers

Reports that describe relationships between health and a given pollutant abound. However, it is difficult, and in some cases impossible, to link a specific source of a pollutant, notably burning coal, to a health effect in question. When I can, I will attempt to define the proportion of the pollutant that is attributable to coal. The linkage issue is a particular issue with regard to oxides of sulfur and nitrogen, and particulates. There are many important sources of these pollutants in addition to burning coal. Therefore in this book I draw on the literature that goes beyond that in which authors limit to coal as the sole source of the pollutant in question. This is rarely possible. Linkage issues are less of a problem for mercury. Burning coal is clearly the most important anthropogenic source for this pollutant.

I recognize that not all chapters will be of interest or pertinent to the needs of all readers. Therefore I sought to make each chapter capable of standing on its own. This explains some of the redundancies that will be evident to those who read the entire book. Hopefully this will not be too irritating to these intrepid souls.

In this book I have employed an organ-systems-based approach to describing the health effects of burning coal rather than a pollutant-based review. This is the format my colleagues and I adopted in *Coal's Assault on Human Health*, available online at http://www.psr.org/coalreport. To minimize bias, whenever possible, I have cited contemporary peer-reviewed medical literature and reports published by governmental agencies such as the US Environmental Protection Agency and the Department of Energy. I hope that this book will provide physicians, other health care providers, policy makers, and concerned citizens with the information they need to make informed choices that affect our health and the future of burning coal.

References

1. Markandaya A, Wilkinson P. Energy and health 2: electricity generation. Lancet 2007;370:979–90.

2. Fumifugium EJ. The Inconvenience of the Aer And Smoak of London. Together with some Remedies Humbly Proposed by J.E. Esq; To His Sacred Majestie and to the Parliament now Assembled. London: Published by His Majesties Command, 1661.

3. Simkhovich BZ, Kleinman MT, Kloner RA. Air pollution and cardiovascular injury epidemiology, toxicology, and mechanisms. J Am Coll Cardiol 2008; 52(9):719–26.

4. Clancy L, Goodman P, Sinclair H, Dockery DW. Effect of air-pollution control on death rates in Dublin, Ireland: an intervention study. Lancet 2002;360(9341): 1210–4.

5. Edison Electric Institute. www.eei.org. Accessed 2008 Sep 1.

6. BP Statistical Review of World Energy. London: British Petroleum, 2010.

7. Study of Hazardous Air Pollutant Emissions from Electric Utility Steam Generating Units—Final Report to Congress EPA publication 453/R-98–004a. Washington DC: EPA, 1998.

8. Goodell J. Big Coal: The Dirty Secret behind America's Energy Future. Boston: Houghton Mifflin, 2006.

9. US Energy Information Administration. Electricity in the United States. Washington DC: Government Printing Office, 2010.

10. US Energy Information Administration. US Coal Consumption. Washington DC: Government Printing Office, 2011.

2
Coal

They digge out of the mountaynes a certayne kinde of blacke stone whiche burne in the fyre like coles.
—Sebastian Münster, 1553 [1]

This quotation is an incomplete but good start at describing the essential features of coal. Coal is not a single, well-defined entity. Coals form a family of extraordinarily complex minerals composed primarily of carbon, but also containing a long list of elements and compounds. Thus there are many different forms of coal whose compositions reflect the conditions under which they were formed. Coals are typically assigned to one of four different ranks, as described in table 2.1.

The different ranks of coal were formed during many geological eras. The largest amount of coal began its journey to the present during the Carboniferous Period, 360 to 290 million years ago. Other coals date to the Permian and Secondary Periods of the Cenozoic Era, 250 to 65 million years ago, as well as the more recent Tertiary Period of the Cenozoic Era, which began about 65 million years ago. These were times when lush forests covered large portions of the existing land masses. When trees fell, they were concentrated in basins where they were covered by water and silt. This kept oxygen from reaching the dead vegetation and prevented the oxidation of the cellulose in the plant matter. This is why there are large amounts of carbon in all of the ranks of coal and why it is a valuable fossil fuel. The threats to health and the environment are due to the release of carbon dioxide and other molecules and elements in the final stages of the life cycle of coal when it is mined, transported, and burned.

Since the coal-forming aggregations of dead vegetation were disbursed widely, there were many different physical and chemical conditions in

Table 2.1
Ranks of coal

Rank	Average CO_2 emission factor, lb/million Btu (kg/GJ) [13]	Carbon,[a] moisture,[b] volatile matter,[b] and ash[b] contents	Heat production, Btu/lb (MJ/kg) [14]
Anthracite	227.4 (98)	86% or more, 4%, 7%, 10%	15,000 (35)
Bituminous	205.3 (88.5)	45–86%, 5%, 38%, 9%	10,500–14,000 (24–33)
Subbituminous	211.9 (91.3)	35–45%, 25%, 35%, 7%	8,200–11,200 (19–26)
Lignite	216.3 (93.2)	25–35%, 38%, 27%, 6%	7,000 (17)

a: Energy Kids: Nonrenewable Coal, US Energy Information Administration. Available at http://www.eia.doe.gov/kids/energy.cfm?page=coal_home.
b: The Science and Technology of Coal, G. E. Dolbear and Associates, Diamond Bar, CA. Available at http://www.coalscience.com.

these deposits. This created large variations in the composition of the coals as they formed. Some were close to volcanoes, a natural source of mercury. Thus mercury and other elements that are the result of vulcanism were added to the organic debris that was destined to become coal. The chemicals in the water that covered these early sites played a critical role in determining final composition of the coal [2]. Seawater was typically rich in sulfur during the coal-forming eras. When it covered these organic deposits, various chemical reactions led to the trapping of sulfur. This accounts for the so-called super-high organic sulfur coal from Guiding, Guizhou, China. The sulfur content of the coal from this deposit may reach 10.5% by weight. At the other end of the sulfur spectrum, the low-sulfur coal from the Powder River Basin in the United States contains less than 1% sulfur by weight. It was covered by freshwater that was low in sulfur as opposed to seawater.

When it is burned, the sulfur in the coal combines with oxygen to form oxides of sulfur, collectively referred to as SO_x, where x refers to the number of atoms of oxygen that have combined with the sulfur. Since SO_x have critical health and environmental effects, the sulfur content of coal is one of the factors that determines its quality. Therefore coals are graded and marketed according to their sulfur content [3].

Low-sulfur coal contains less than 0.6 pounds of sulfur per million British thermal units (Btu) of heat content, a measure of the amount of heat produced per unit weight of coal when it is burned. Medium-sulfur coal contains between 0.61 and 1.67 pounds of sulfur per million Btu, and high-sulfur coal contains more than 1.67 pounds of sulfur per million Btu.

As time passed, many layers of sediments were deposited over these nascent coal beds. Different combinations of time, heat, and pressure contributed to the characteristics of the different ranks of coal. There are four principal types, or ranks, of coal. In order of increasing carbon content, these are lignite, subbituminous, bituminous, and anthracite coal. Lignite, sometimes called "brown coal," is the youngest, most recently formed, and lowest quality or rank of coal, where quality is determined by the amount of heat produced when it is burned. Like subbituminous and bituminous coal, lignite is a sedimentary rock. As shown in table 2.1, lignite, like other ranks of coal, contains a substantial amount of moisture and volatile matter. As the quality of the coal increases, the moisture content decreases. Coals also contain a volatile component, made up of various hydrocarbons, including methane. This volatile fraction is highest in subbituminous and bituminous coals. It is this volatile component of coal that forms soot or black carbon when low- and mid-ranked coals are burned under suboptimal conditions. Soot formation is minimized in modern boilers where conditions can be optimized to ensure complete combustion. Anthracite coal, the rarest of the ranks of coal, has the highest carbon content, and the smallest amount of moisture and volatile matter. It is a metamorphic rock, and was formed from a lower rank of coal as additional heat and pressure drove water and organic compounds from the anthracite precursor. In the United States most anthracite deposits are found in what is now northeastern Pennsylvania where folding of the earth's crust formed the Appalachian Mountains. Anthracite coal yields the most ash, on a weight basis. This is due to its low moisture and volatile matter content, leaving little room for anything else other than ash and carbon.

There are other terms that indicate how coals are used. For example, steam coal refers to coal that is used to produce steam—most of which is used to generate electricity. Typically this is bituminous coal. Other coals that are rich in volatile components, as shown in table 2.1, are used to produce coke. Most coke is produced by heating bituminous coal that has a low-ash and low-sulfur content. This removes the volatile components by distillation and leaves behind a porous, high-carbon coke. Coke is used in smelters to remove the oxygen chemically bonded to iron in iron ores in order to make iron and steel.

Perhaps the most telling information concerning the complexity of coals came from rigorous analyses of the emissions created when coal is burned [4]. A 1998 EPA report, which was mandated by the Clean Air Act, identified 67 hazardous air pollutants in the emissions from utilities in operation during the early 1990s. These emissions included hydrofluoric acid, mercury, arsenic, beryllium, cadmium, chromium, lead, manganese, mercury, nickel, uranium, and thorium, as shown in table 3.2. Because of the nature of the Clean Air Act, pollutants, such as oxides of nitrogen and sulfur as well as particulate matter, were excluded from this survey.

World coal reserves are enormous. Proven reserves total just over 908 billion tons [5]. These reserves are those that can be mined now and in the future, given present-day technology and expected economic conditions. Just over 28% of proven reserves are in the United States, 13.5% are in China, and just over 18.5% are in the Russian Federation [6]. Estimates of additional deposits in the United States, made using conservative, nonspeculative assumptions, could raise US deposits by as much as 1,200,000 million tons. Similar estimates are not available for China; however, estimates of the additional deposits of bituminous coal in the Russian Federation exceed 200,000 million tons.

These estimates may be overly optimistic. In 2002 Ruppert and colleagues evaluated the US Geological Survey's assessment of US coal resources [7]. They concluded that the coal that can actually be recovered, given coal quality considerations, largely related to sulfur content, land use, other regulations, and technological considerations, place doubt on the survey's conclusion. For example, they conclude that only one-tenth of the coal reserves in the Illinois Basin is economically recoverable. Glustrom, in a posting on the website of Clean Energy Action, an advocacy organization whose mission is to promote clean energy and alternatives to fossil fuels, concludes that economically recoverable coal reserves in the United States may only be sufficient to last twenty to thirty years [8].

Despite these predictions, the growth of coal as an energy source exceeds all others: at current rates of use, the World Energy Council concluded in their 2010 report that coal supplies should last at least 150 years [6]. By any measure, it seems likely that coal is certain to be with us for a very long time. As coal prices rise, more and more coal will be economically recoverable. Goodell and others refer to the United States as the "Saudi Arabia of Coal" [9].

According to the US Energy Information Administration, 1,406 mines produced just over 1 billion tons of coal in 22 states in 2009 [10]. This represented an 8.3% reduction in the amount mined, compared to 2008. Almost 50% of all coal mined in the United States comes from just 10 deposits [10]. About one-third of this coal was mined in the Appalachian region. The three leading coal states are Kentucky, West Virginia, and Pennsylvania.

Despite this decline in coal production, mirrored by similar worldwide reductions associated with the recession, the International Energy Organization projects a 56% increase (from 132 quadrillion Btu to 206 quadrillion Btu) in coal consumption in the interval between 2007 and 2035 [11]. Ninety-five percent of this increase is expected to occur in countries that are not a part of the Organization for Economic Co-operation and Development (OECD), mainly China and India. Annual energy use rates are expected to increase by 5% per year in India and 5.8% in China. By contrast, coal consumption in OECD countries is expected to decrease by about 10% in the interval between 2007 and 2025 and remain relatively unchanged for the ensuing decade. This projection is made despite an anticipated growth in coal use in the United States of about 11%, from 22.7 to 25.1 quadrillion Btu. Decreases in coal use are the result of several factors including an increase in the use of natural gas and current economic conditions.

Even though coal deposits are found in many locations throughout the world, the international coal trade is growing in importance. Most of the importation of coal occurs in Asia. This region already accounts for almost 60% of all coal imports (12.6 quadrillion Btu) and is expected to increase to 70% by 2035 [11]. The majority of this growth is expected to occur during the next decade and then taper off as China expands its capacity to move its own coal from coal-rich regions to where it is used. This growth and changes in Chinese coal use are due to several factors, including the expanding economy, differences in the cost of transporting coal from inland mines to coastal manufacturing sites, and the fact that much of China's coal contains large amounts of sulfur, compared to the low-sulfur coal produced in parts of the United States, such as the Powder River Basin. In 2008 the major suppliers of coal targeted for the production of electricity were Indonesia, Australia, South America (mainly Colombia), and southern Africa (mainly South Africa) [11].

The large number of coal states and the growing importance of the exportation of coal accounts, in part, for the influence of coal lobbies in Washington, DC. According to data released by the Federal Election Commission on October 25, 2010, three of the top five political action committees (PACs) in the energy and natural resources sector have coal-based interests. These three PACs reported expenditures of $492,000, 60% of which went to Republican candidates [12]. As the result of the *US Supreme Court Decision in Citizens United* v. *Federal Election Commission* granting First Amendment rights to corporations, the total amount of coal money spent on lobbying can no longer be determined with certainty. It is likely to be much more than the amount reported by PACs.

References

1. Münster S. A treatyse of the newe India, with other new found landes and ilandes . . . (trans Eden R). London, 1553.

2. Chou C-L. Geologic factors affecting the abundance, distribution, and speciation of sulfur in coals. In: Yang Q, ed. Proceedings of the 30th International Geological Congress. Zeist, Netherlands: VSP BV, 1997.

3. US Energy Information Administration. US Coal Reserves: An Update by Heat and Sulfur Content. Washington DC: US Department of Energy, 1993.

4. Study of Hazardous Air Pollutant Emissions from Electric Utility Steam Generating Units—Final Report to Congress. EPA publication 453/R-98–004a. Washington DC: EPA, 1998.

5. BP Statistical Review of World Energy. London: British Petroleum, 2010.

6. World Energy Council. Survey of Energy Resources 2007. Available at http://www worldenergy.org. Accessed 2010.

7. Office of Management and Budget OoIaRA. Informing Regulatory Decisions: 2003 Report to Congress on the Costs and Benefits of Federal Regulations and Unfunded Mandates on State, Local, and Tribal Entities. Washington DC: Government Printing Office, 2003.

8. Council of Economic Advisors EOotP. The Economic Case for Health Care Reform. Washington DC: Government Printing Office, 2009.

9. Goodell J. Big Coal: The Dirty Secret behind America's Energy Future. Boston: Houghton Mifflin, 2006.

10. US Energy Information Administration. Annual Coal Report 2009. Washington DC: Government Printing Office, 2010.

11. US Energy Information Administration. International Energy Outlook. Washington DC: Government Printing Office, 2010.

12. PAC Contributions to Federal Candidates: Energy and Natural Resources Sector. Center for Responsive Politics. Available at opensecrets.org. Accessed 2010.

13. Hong BD, Slatick ER. Carbon Dioxide Emission Factors for Coal. Washington DC: US Energy Information Administration, 1994DOE/EIA-0121(94/Q1).

14. Encyclopedia Britannica Online. Available at http://www Britannica.com. Accessed 2010.

3

The Pollutants

Overview of Exposure to Hazardous Air Pollutants

Although carbon is the principal element found in coal, coal is actually a complex mixture of numerous elements and compounds. Many of these are released into the atmosphere as hazardous air pollutants when coal is burned and many of them are harmful to health. Studies done by the EPA have identified some 67 hazardous pollutants that are discharged into the air by coal-fired utilities [1].

An individual or a population must be exposed to a pollutant before it can exert its effect. Although this is axiomatic, numerous critical factors play complex roles between the time that a pollutant is released into the environment and the time that an exposure occurs. In the case of airborne pollutants, the concentration in air, usually expressed in terms of mass per unit volume, such as $\mu g/m^3$ (microgram per cubic meter), is perhaps the most important of these. Concentration is, in turn, affected by other factors, such as the rate of emission of the pollutant, and local weather and wind conditions. Daily and seasonal fluctuations in these variables cause additional complexity. Some of the emission rate variability depends on the demand for electricity. During hot weather, when air-conditioning use peaks, it is common to find concomitant peaks in electricity usage. To meet power demands, additional generating facilities with less efficient pollution control technologies may be brought on line to meet demand.

Once released, some pollutants stay in the atmosphere longer than others. For example, very small particles, those with a diameter of 2.5 μm (microns) or less, have long atmospheric lifetimes and may therefore travel substantial distances from the site of emission to the site of exposure. This widespread dispersion creates a large area wherein an exposure may occur. Other pollutants, such as mercury, are deposited closer to the site of emission, causing local hot spots.

Macro- and micro-environmental factors affect exposures. Some of these include whether an individual is outdoors where pollution levels are generally highest, or indoors, where levels are typically lower. Activity levels are also important. With high levels of activity, the amount of air breathed per unit time increases, leading to corresponding increases in exposure to airborne pollutants. Children breathe more air per unit body weight than adults and they are usually more active than adults. These factors make them more susceptible to the adverse effects of air pollution.

The quality of the air at a given location is an important determinant of health. It is the result of the complex interaction of all of the factors mentioned above. In order to aid individuals and communities in their efforts to deal with the effects of air pollution, cooperative efforts between the EPA, state, and local agencies have established Enviroflash, an online resource that provides health-oriented up-to-date air quality information. This information is available at http://www.enviroflash.info. During temperature peaks that occur during the summer, air pollution advisories that involve major metropolitan areas in the United States can be found on this website. During the air pollution advisory, sensitive populations are urged to minimize physical activity and to remain indoors.

Hazardous Air Pollutants and the Clean Air Act

The Air Pollution Control Act of 1955 marked the beginning of federal involvement in issues relevant to air pollution [2]. The ensuing report drew attention to the fact that air pollution was potentially injurious to health. It left the control of pollutants up to the states, but authorized the federal government to conduct additional research and disseminate information. The Air Quality Act of 1967 expanded federal studies of air pollution to include emission inventories, monitoring and control technologies.

The focus changed from research to action as a result of the data yielded by these two acts and the increasing realization that air pollution posed serious threats to health and the environment [2]. This led to the adoption of the National Environmental Policy Act, the legislation that created the US Environmental Protection Agency (EPA), and significant amendments to the Clean Air Act that was originally passed in 1963. This EPA was created on May 2, 1971, with a mission to " . . . protect human health and to safeguard the natural environment—air, water and land—upon which life depends." Currently there are four major regulatory provisions of the Act: National Ambient Air Quality Standards (frequently abbreviated as NAAQS, pronounced nacks), State Implementation Plans, New Source Performance Standards, and National Emission Standards for Hazardous Air Pollutants. The EPA is also the home of one of the most bewildering arrays of terms, abbreviations and acronyms. For help, see the glossary at the end of this book and the EPA Terminology Services website, available at http://iaspub.epa.gov/sor_internet/registry/termreg/home/overview/home. Ambient air quality standards define a primary standard that is designed to preserve health, and a secondary standard that is designed to preserve the environment for each of six pollutants that were determined to pose the greatest threats to health. These six pollutants are known as criteria pollutants because of the primary and secondary standards or criteria associated with each. These pollutants are carbon monoxide, lead, particulate matter (now subdivided by particle size), ozone, nitrogen dioxide, and sulfur dioxide. Current standards are shown in table 3.1. Other hazardous air pollutants are included in table 3.2 and discussed below.

The Clean Air Act has been amended two more times. The 1977 amendments were designed to identify regions of the country where air quality was deteriorating or where air quality standards had not been realized. These are referred to as sites of nonattainment. The 1990 amendments to the Act increased the authority of the EPA to deal effectively with air pollution, especially oxides of sulfur and oxides of nitrogen, the pollutants that cause acid rain (in many publications these oxides are commonly referred to as SO_x and NO_x).

In 1991, as a part of its congressional charge, the EPA began to perform a systematic evaluation of the risks associated with burning coal to produce electricity [1]. The charge included this specific language: ". . . perform a study of the hazards of public health reasonably anticipated to occur as a result of emissions by electric utility steam generating units of . . . [hazardous air pollutants]. . . ."

Table 3.1
National ambient air quality standards

Pollutant	Primary standards		Secondary standards	
	Level	Averaging time	Level	Averaging time
Carbon monoxide	9 ppm (10 mg/m³)	8 hour[a]	None	
	35 ppm (40 mg/m³)	1 hour[a]		
Lead	0.15 µg/m³	Rolling 3-month average	Same as primary	
	1.5 µg/m³	Quarterly average		
Nitrogen dioxide	53 ppb	Annual arithmetic average	Same as primary	
	100 ppb	1 hour	None	
Particulate, PM₁₀	150 µg/m³	24 hour[b]	Same as primary	
Particulate, PM₂.₅	15 µg/m³	Annual arithmetic average	Same as primary	
	35 µg/m³	24 hour	Same as primary	
Ozone	0.075 ppm; 2008 standard	8 hour	Same as primary	
	0.12 ppm	1 hour	Same as primary	
Sulfur dioxide	0.03 ppm	Annual arithmetic average	Same as primary	
	0.14 ppm	24 hour[a]	0.5 ppm	3 hour
	75 ppb	1 hour[c]	None	

Note: ppm and ppb = parts per million and billion, respectively. Per the EPA NAAQS website, primary standards set limits to protect public health, including the health of "sensitive" populations such as asthmatics, children, and the elderly. Secondary standards set limits to protect public welfare, including protection against decreased visibility, damage to animals, crops, vegetation, and buildings.
a. Not to be exceeded more than once per year.
b. Not to be exceeded more than once per year on average over 3 years.
c. The final rule was signed on June 2, 2010. To attain this standard, the 3-year average of the 99th percentile of the daily maximum 1-hour average at each monitor within an area must not exceed 75 ppb.

Table 3.2
Emissions of priority hazardous air pollutants [1]

HAP	1994 Emissions (tons per year)	Cancer risk[c]	Toxicity[b]
Arsenic (As)	56	3×10^{-6}	Long-term ingestion of small amounts may affect skin (hyperpigmentation, corns, and warts), damage peripheral nerves (painful sensation of "pins and needles"), and increase risk of cancer of urinary bladder and lung.
Beryllium (Be)	7.9	3×10^{-7}	In comparison with other elements (lead, chromium) Be exposure is insignificant. Most ingested Be is eliminated in the feces. Inhaled Be is more persistent. Inhalation of Be compounds (greater than 1 mg [milligram] per cubic meter) may cause acute or chronic lung disease. The average Be concentration in US urban air is 0.2 ng per cubic meter; 1 ng = 1 billionth of a gram.
Cadmium (Cd)	3.2	2×10^{-7}	Cd accumulates in shellfish (observe fishing advisories), organ meats, lettuce, spinach, potatoes, grains, peanuts, soybeans, sunflower seeds, and tobacco. Inhalation of low levels of Cd for years or consumption of food with elevated Cd may cause kidney disease or fragile bones.
Chromium (Cr)	62	2×10^{-6}	Cr is a known carcinogen. Concentrations in air are typically less than 2% of those that cause respiratory problems in Cr workers. Avoid tobacco smoke and older pressure-treated lumber to minimize exposure.
Lead (Pb)	62	NA	Children are more vulnerable that adults. Neurological problems include encephalopathy (global brain dysfunction) producing behavioral and cognitive deficits, damage to peripheral nerves, anemia, and kidney damage.
Manganese (Mn)	168	NA	Major exposure comes via consumption of large amounts of grains, beans, nuts, tea, and nutritional supplements. Small amounts are inhaled. Accumulation in the brain, particularly in patients with liver disease, may cause symptoms similar to Parkinson's disease.
Mercury (Hg)	51	NA	See the text for detailed discussion.
Nickel (Ni)	52	4×10^{-7}	There are multiple routes of exposure including inhalation. Food is the major source. The average concentration in air is 2.2 ng/m³ and is attached to small particles. Concentrations found to cause cancer were 100,000 to 1 million times greater than that commonly in air, and 10% to 20% if people are allergic to Ni.

Table 3.2
(continued)

HAP	1994 Emissions (tons per year)	Cancer risk[c]	Toxicity[b]
Hydrogen chloride (HCl)	134,000	NA	Rain removes from atmosphere, limiting exposure from HCl released into the air.
Hydrogen fluoride (HF)	23,000	NA	Rain removes from atmosphere, limiting exposure from HF released into the air, low concentrations of fluorine harden teeth and bones.
Acrolein	27	NA	Inhalation causes irritation of nasal mucosa or other parts of the respiratory tract. Outdoor air concentrations range between 0.5 and 3.2 ppb (parts per billion). Minimum risk levels for chronic duration inhalation are not available and are about 3 ppb for exposures of less than 14 days. Environmental tobacco smoke is the major cause of exposure.
Dioxins[a]	0.00020	5×10^{-8}	Dioxins are probably carcinogens. They may cause a variety of skin problems, including chloracne. Type 2 diabetes and other endocrine disorders have been attributed to dioxin exposure.
Formaldehyde	29		Formaldehyde decomposes to formic acid and carbon monoxide within a day. Air concentrations in the highest areas are 10 to 20 ppb. Many home products release formaldehyde, and indoor air concentrations are usually higher than in outdoor air. Formaldehyde is an irritant and is dangerous to life at a concentration of 20 ppm. It is likely to be a carcinogen.

a. Dioxin emissions are the summation of dioxin equivalents for each member of this family relative to 2,3,7,8-tetrachlorodibenzo-*p*-dioxin

b. Toxicity information was obtained from the Agency for Toxic Substances and Disease Registry ToxGuides™. Available at: http://www.atsdr.cdc.gov/toxguides/index.asp, Public Health Statements; http://www.atsdr.cdc.gov/PHS/Index.asp, or ToxFAQs™; http://www.atsdr.cdc.gov/toxfaqs/index.asp.

c. Cancer risk = highest cancer risk for maximally exposed individual due to inhalation of the hazardous air pollutant for 70 years at the highest presumed concentration. For details of modeling, see [1], section 6.1.1 of and the health risks sections of the appendixes in volume 2.

Five years later, the Agency published its initial report, followed by the final version titled, "Study of Hazardous Air Pollutant Emissions from Electric Utility Steam Generating Units—Final Report to Congress." This two-volume opus remains as one of the most definitive sources of data about the emissions associated with burning coal to produce electricity [1]. There were 684 utility plants that informed that study. A hazardous air pollutant emissions testing program was established at 52 of these plants that were selected because they were representative of the industry as a whole. The testing program identified 67 pollutants with a potential for emission by utilities. These pollutants underwent an additional evaluation to estimate the risks to individuals who were likely to have the largest exposures. The EPA scientists used a conservative Human Exposure Model to identify a subset of these pollutants that posed the greatest risk to health. The subset of these pollutants, whose exposure was primarily by inhalation, was augmented by others whose exposure was by non-inhalational routes if the pollutant was toxic, persistent, tended to bioaccumulate, was emitted in large quantities, or was radioactive. A total of 14 pollutants was identified and became the focus of the final report. These, minus radionuclides (discussed below), along with 1994 emission projections and their health effects, are shown in table 3.2. Since they are regulated under other provisions of the Clean Air Act, oxides of sulfur, oxides of nitrogen, and particulates were not considered in this analysis. These pollutants form the core of the criteria air pollutants. Because of their critical role in the production of the adverse health effects associated with burning coal, criteria air pollutants are discussed in the next section of this chapter.

A separate modeling strategy was used to estimate the risks posed by the emission of radionuclides. The nuclides of greatest concern are uranium and thorium plus the products of their radioactive decay. The modeling predicted that the highest multipathway (inhalation, ingestion, etc.) exposure to radiation would lead to an absorbed dose of about 1.5 millirems per year. This is about 1.5% of a reasonable estimate for the natural background radiation exposure (note: background radiation exposures vary substantially depending on a number of factors). The report predicted that this radionuclide exposure would translate into about 0.3 cancer deaths per year for inhabitants living within 50 kilometers (about 30 miles) of each utility.

In 1986, in legislation unrelated to the Clean Air Act, Congress created the Toxics Release Inventory based on the premise that the public had a right to information about the release of over 600 chemicals and elements into the environment. The TRI is now available at http://www.epa.gov/tri.

Arsenic

Of the compounds listed in table 3.2, the health risks associated with arsenic warrant a more detailed discussion. Arsenic pollution is a concern in virtually all aspects of coal production and use ranging from leaching from mine tailings, air emissions during combustion, and, more recently, as a component of coal combustion waste, namely coal ash.

Of the pollutants in the table, arsenic may be the most familiar to the general population. The familiarity of this poison is exemplified by the farcical black comedy "Arsenic and Old Lace" in which two elderly spinsters poison male boarders with elderberry wine that contains arsenic.

Arsenic is a class A carcinogen, meaning it is known to cause cancer in humans [3]. The inhalation of arsenic is associated with an increased risk for the development of lung cancer, and ingestion, typically in drinking water, the principal means of human exposure, is associated with an increased risk of skin cancers (excluding melanomas), and carcinomas of the urinary bladder, liver, and lung. Arsenic is found in particulate matter that has an aerodynamic diameter of 2.5 μm or less. Other trace metals, including antimony, beryllium, cadmium, cobalt, chromium, iron, lead, manganese, mercury, nickel, selenium, and zinc, are also found in these small particles [4]. The concentration of arsenic in the air varies by location, and ranges between 0.5 and 7.2 ng/m^3 (nanograms per cubic meter) [4]. Particulate pollution is very toxic, as discussed elsewhere in this and other chapters. It is not clear whether arsenic-containing particles are more or less toxic than other particulates. At the present time the EPA classifies particulates by size and not by composition.

There are many assessments of the risks associated with arsenic. In the EPA report on HAPS, the Agency estimated that the lung cancer risk associated with arsenic inhalation was about one in a million [1]. This estimate was based on the presumed air concentration of 0.0002 μg/m^3

over a lifetime of 70 years [1]. In another report, the State of California Air Resources Board estimated that between 16 and 40 Californians died each year from arsenic-induced cancers [5].

Based on these assessments, it is reasonable to conclude that the risks posed by air emissions of arsenic are relatively small, particularly when considered alongside the risks posed by oxides of sulfur, nitrogen, mercury, and particulates of all types. This does not mean that arsenic emissions should not be curtailed, to the extent possible. The greatest risk posed by arsenic may be that which is associated with the disposal of coal combustion wastes, as discussed in chapter 4.

Criteria Air Pollutants

As discussed in the section on the historical evolution of air pollution controls, the term "criteria air pollutants" arose as a result of amendments to the Clean Air Act provisions that required the EPA to focus on the health and welfare effects of six pollutants known to be hazardous to human health. The criteria pollutants most relevant to burning coal are sulfur dioxide, nitrogen dioxide, ozone, and particulate matter. Carbon monoxide and lead are also criteria pollutants. Lead was included at the time the list was formulated because of the large amount emitted by burning leaded gasoline. The Clean Air Act mandates periodic revision of these standards as new knowledge emerges. The political landscape all too often trumps science in EPA decision-making, as discussed in chapter 14.

Oxides of Sulfur

Almost all sulfur in coal forms sulfur dioxide when it is burned, regardless of its initial chemical form. Smaller amounts of sulfur trioxide are also produced. Therefore the term SO_x is used frequently to refer to all oxides of sulfur. SO_x that are present in the air as gases include sulfur dioxide, sulfur trioxide, and gaseous sulfuric acid. These compounds are toxic in and of themselves. In addition sulfur oxides make an important contribution to the formation of secondary particulate matter, as explained below. Burning coal to produce electricity is by far the leading source of US sulfur dioxide emissions. The EPA Acid Rain Program has been effective in reducing these emissions from 13.1 million tons in 1998 to 5.7 million tons in 2009 as shown in figure 3.1.

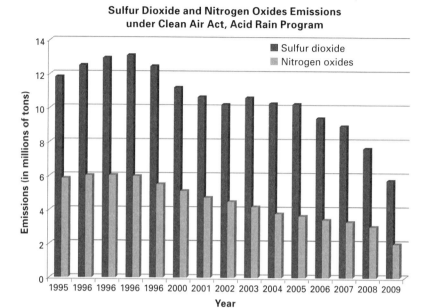

Figure 3.1
US emissions of oxides of sulfur and nitrogen from 1995 to 2009. Decreases are attributable to the EPA Acid Rain Program of the Clean Air Act. Data available at http://camddataandmaps.epa.gov/gdm/index.cfm.

Sulfur dioxide is the most important of the oxides of sulfur, in terms of the mass emitted and health effects. It is a colorless gas at room temperature and pressure. It dissolves rapidly and completely in water or aqueous solutions, such as the moist linings of the nose and the upper airway. It is easily identified because of its highly characteristic, strong, pungent, and unpleasant smell. In the United States, almost all emissions of oxides of sulfur are the result of burning fossil fuels, mainly coal, by electrical utilities, and industrial fossil fuel use [6]. These two categories account for about 95% of anthropogenic emissions of the oxides of sulfur. Electrical utilities and industrial boilers are referred to as point sources because they are discrete and immobile. The remaining sulfurous emissions arise from the transportation industry and natural sources such as volcanoes and wildfires. Worldwide, human activity is thought to be responsible for about 99% of the sulfur dioxide in the air. Small amounts are released by volcanoes where the rotten egg smell of hydrogen sulfide is often present as well. In addition small amounts of sulfur dioxide are formed secondarily from naturally occurring sulfur-containing compounds in chemical reactions that use sunlight as an energy source [6]. These small distributed sources of SO_2 formation are found in marine environments along the east and west coasts of the United States. Their contribution to the total atmospheric concentration of sulfur dioxide is minute.

Once released into the atmosphere, sulfur dioxide forms sulfuric acid via a series of chemical reactions. Coal-fired power plants release large amounts of both of these gases. Sulfuric acid is the major component of acid rain. The direct effects of sulfur dioxide as a powerful irritant, its propensity to cause acid rain and to form particulates, and the enormity of the amounts released into the atmosphere, make this gas a leading cause of the adverse health effects of burning coal [7].

The impact of the oxides of sulfur was identified and quantified as "external damages" in the National Research Council report on the *Hidden Costs of Energy* caused by burning coal [7]. External damages are the result of converting the adverse effects of a pollutant, on morbidity, premature mortality, crop yields, and so forth, into a dollar cost. The costs attributed to sulfur dioxide were projected to be on the order of 3.8 cents/kWh (per kilowatt hour) of electricity [7]. Because of uncertainties in the data and various assumptions used to make these projections, such as population density around a given plant and adequacy of pollution control devices, the range of the calculated damage estimates was substantial. The lowest estimate, at the 5th percentile, was 0.24 cents/kWh and the highest, at the 95th percentile, was 11.9 cents/kWh. Not surprisingly, the oldest plants with the fewest pollution control devices were the most important sources of this pollutant. Those built before 1950 typically emitted 20.58 lb/MWh (pounds per million watt hours). of electricity generated, whereas those built after 1990 and equipped with modern pollution control devices that perform flue gas desulfurization, typically emitted 3.88 lb/MWh.

Sulfur emitted by point sources, such as coal plants, returns to the earth in liquid or dry forms. Liquid sulfur compounds are a major component of acid rain. Dry forms include the secondary particles that are formed when sulfurous compounds in the air react with other airborne compounds to form small particles. Typically these particles have an aerodynamic diameter that is less than or equal to 2.5 μm. Sulfur-containing particles form a major portion of those in that size category, particularly in the eastern part of the United States.

Much of the sulfur emitted by point sources, such as coal-fired boilers, is deposited relatively close to the source. This is illustrated clearly in part A of figure 3.2, which depicts wet and dry deposition of SO_x in the Northern Hemisphere [8]. High deposition rates are shown in the United States east of the Mississippi River, eastern Europe, southern and northeastern India, Korea, Japan, and the eastern and northeastern parts of China. These are all regions of heavy industrialization where air pollution poses significant health and environmental problems. There is substantial congruity between sulfur deposition and the location of the coal plants in the United States that are associated with the greatest external costs, largely in the form of adverse health outcomes, as shown in figure 13.2 [7].

In 2007 there were approximately 500 sites in the United States where the atmospheric concentration of sulfur dioxide was measured, as required by the EPA. Data from these sites show that the highest sulfur dioxide concentrations in the air are present in the Ohio River Valley [6]. Steubenville, Ohio, bears the dubious distinction of having the highest known mean concentration in its air of any city in the United States. The concentration varies substantially at different times of the day, being highest during the day and during the summer [6]. These are the times when the consumption of electricity is the highest. Although atmospheric conditions, such as air flow and air current patterns, play a role in determining sulfur dioxide concentrations on an hour-to-hour basis, coal-produced electrical power generation rates, which are themselves subject to diurnal and seasonal variations, are critical factors.

Oxides of Nitrogen

NO_x, is a generic term used in many publications and reports that refers to chemical combinations of nitrogen, the most abundant gas in the atmosphere, with oxygen, the second most abundant atmospheric gas. Nitrogen oxides are formed during the combustion of fossil fuels at high temperatures. It takes a lot of energy to force the combination of these two elements. The high temperatures that are critical to efficient

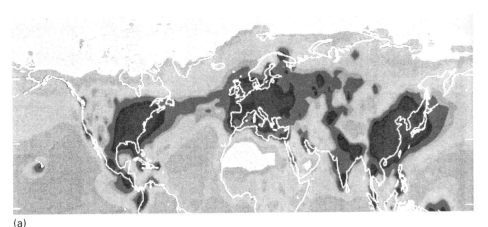

(a)

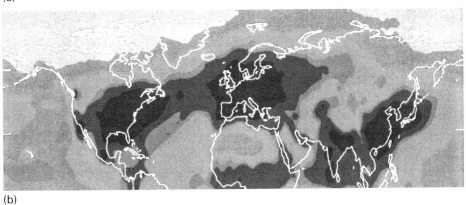

(b)

Figure 3.2
Northern Hemispheric depositions of oxides of sulfur and oxides of nitrogen.
(a) Deposition rates for sulfur oxides and (b) deposition rates for oxides of
nitrogen. Darker areas indicate higher deposition rates [8]. Note: this figure is
adapted from the full color versions in the original publication. Reprinted with
permission from Environmental Science & Technology. Copyright 2002 American Chemical Society.

combustion processes are the key to the formation of nitrogen oxides. At the exact moment of combustion, nitric oxide is the most abundant chemical in the family of nitrogen oxides. However, nitric oxide is chemically unstable and combines spontaneously with oxygen in the atmosphere to form nitrogen dioxide, a reddish-brown gas with a very strong and pungent odor. Nitrogen dioxide is responsible, in part, for the brownish haze that is characteristic of ground-level smog.

There are two major mechanisms for the formation of the oxides of nitrogen. These create the thermal and fuel oxides [9]. Thermal oxides of nitrogen are those formed by the direct, high-temperature, oxidation of the nitrogen present in the atmosphere. This reaction increases in importance at temperatures in excess of 2,800 degrees Fahrenheit. Fuel oxides of nitrogen are formed by the chemical reaction of nitrogenous compounds present in the fuel itself with oxygen in the air during combustion. Predictably, fuel oxide production increases as the nitrogen content of the fuel increases and accounts for as much as 80% of the oxides of nitrogen formed by burning coal. Burning natural gas and liquid fuels typically form smaller amounts of fuel oxides of nitrogen than formed by burning coal.

The definition of oxides of nitrogen that is contained in the National Ambient Air Quality Standards legislation is different from that which is commonly used in other discussions of air pollution. The conventional definition limits this category to nitric oxide and nitrogen dioxide. The air quality legislation expands this definition to include *all* forms of oxidized nitrogen. For most practical purposes, and in this chapter, the term oxides of nitrogen will refer mainly to nitric oxide and nitrogen dioxide. A more complete list of the oxides of nitrogen, as required by air quality legislation, is presented in the EPA Integrated Science Assessment for Oxides of Nitrogen [9,10].

EarthTrends reports that worldwide emissions of oxides of nitrogen reached nearly 140 million tons in 2000 (available at http://earthtrends.wri.org, searchable database results). North American emissions trailed those of Asia (including China) and Europe. Panel b of figure 3.2 depicts worldwide depositions of nitrogen oxides. These resemble closely the deposition of the oxides of sulfur, as shown in panel a of the figure. In 2005 gasoline and diesel engines were the most important sources of nitrogen oxide emissions in the United States. These sources are close to the ground and mobile, so the oxides they form are more highly concentrated and travel shorter distances than those formed in coal-burning boilers. Electrical utilities, the second-ranked source of nitrogen oxides, emitted 3.78 million tons in 2005. These emissions come from stationary point sources and are usually discharged higher into the atmosphere by smoke stacks where they are diluted by winds and travel longer distances than those emitted by the transportation industry.

There is enormous variability in the spatial and temporal distribution of the oxides of nitrogen and the compounds they produce as the result of chemical reactions in the atmosphere [10]. Inorganic compounds (those that do not contain carbon atoms) include nitrous acid (HONO), nitric acid (HNO_3), pernitric acid (HO_2NO_2), and particulate nitrate (PNO_3^-). In addition to serving as potent irritants, these inorganic compounds are important contributors to acid rain. There is also a large number of organic compounds that are a part of the greater family of nitrogen oxides. Additional details are present in the EPA Integrated Science Assessment for Oxides of Nitrogen [10].

Not all the news is bad. Under provisions of the Acid Rain Program of the Clean Air Act, the emissions of oxides of nitrogen have declined substantially in recent years, as shown in figure 3.1. Emissions in 2009 were approximately one-third of those reported in 1997 as shown in figure 3.1.

Ozone

A great deal of the importance of the oxides of nitrogen stems from their role in the production of ground-level ozone, a major component of smog. It is important to distinguish ground-level ozone from stratospheric-ozone. Stratospheric ozone protects health and the environment by blocking ultraviolet rays from the sun. Ground-level ozone is a potent irritant and an important cause of respiratory disease.

Ozone is a light blue gas with a pungent odor that is similar to that of chlorine. It is formed in the atmosphere by natural processes, such as lightning, as well as those that are the consequence of human activity. These reactions are very complex and depend on many factors, including altitude. Ozone in the troposphere, the layer of the atmosphere that is close to the ground, is highly toxic and a threat to health. Ozone in the stratosphere, the layer of the atmosphere directly above the troposphere, protects us from ultraviolet rays from the sun. This explains the origin of the phrase, "Ozone: good up high, bad nearby."

Ozone is formed in the stratosphere from molecular oxygen [11]. Ultraviolet light from the sun splits molecular oxygen into two atoms of oxygen in the first reaction that leads to the formation of ozone. These highly reactive single atoms of oxygen combine with molecular oxygen to form ozone. Sunlight also splits ozone into molecular oxygen and single atoms of oxygen. In addition to forming ozone, these single atoms of oxygen also attack ozone molecules to form two molecules of molecular oxygen. Thus there is an equilibrium concentration of ozone due to its constant formation and degradation. The heat generated via these reactions, and the associated trapping of the energy from the sunlight that drives them, warms the upper portions of the stratosphere. Without this blockade or absorption of ultraviolet light, it is unlikely that life as we know it could exist.

The chemical composition of the lower portions of the atmosphere (the troposphere) and the reactions that form ozone in this layer are quite different from those in the upper layers of the atmosphere (the stratosphere) [11]. In the troposphere and particularly in the layer closest to the surface of the earth, the planetary boundary layer, the reactive oxygen atoms that combine with molecular oxygen to form ozone come from nitrogen dioxide. The reaction sequence is

NO_2 + sunlight → NO + reactive oxygen molecule

reactive oxygen molecule + O_2 → O_3

In the 1950s atmospheric chemists determined that this reaction could not generate the concentrations of ozone that were actually present in the atmosphere. Their research led to the discovery that volatile (easily vaporized) organic compounds (VOCs) played a critical role. In a highly simplified form, the reactions that generate ground-level or tropospheric ozone can be written as follows:

NO_x + VOC + ultraviolet light → O_3 + other products

There are many volatile organic compounds, and thus hundreds of chemical reactions that are summarized by the above equation. In urban areas, carbon monoxide and a wide variety of carbon-containing compounds contribute to the total concentration of these organic molecules.

In rural areas, methane and carbon monoxide are the most important of the volatile organic compounds (for an account of methane sources, see chapter 7). The ratio of nitrogen dioxide to volatile organic compounds and weather conditions all interact to determine the rate of ozone formation. Ground-level ozone concentrations are typically highest on hot summer days, when the physical and atmospheric conditions are the most favorable for ozone formation.

According to the EPA, almost 16 million tons of these volatile compounds were released into the atmosphere in 2005 (available at http://www.epa.gov/air/emissions/voc.htm). Solvent use, including paints, on-road vehicles, off-road equipment, industrial processes, and miscellaneous sources including office copiers and printers were the major contributors to this total. In that same year nitrogen dioxide emissions totaled about 18.3 million tons.

Particulate Matter

Particulate matter is one of the most important forms of air pollution, particularly in terms of its worldwide impact on health. From this global perspective, particulate matter, in the form of indoor smoke from burning solid fuels, such as coal, wood, and dung, ranks eighth among the top 20 burden-of-disease risk factors [12]. Particulate matter accounted for about 2.6% of all disability-adjusted life years as reported by the World Health Organization in 2002. This burden of disease is borne primarily by those who live in developing countries. Particulate matter is also a major contributor to urban pollution. In the United States, generating electricity accounted for approximately 11.4% or 515,000 tons of the estimated total of 4.48 million tons of small diameter particulates discharged into the atmosphere in 2005 (available at http://www.epa.gov/air/emissions/pm.htm).

Particulate matter is not a single entity with a defined composition, a fact that creates many problems in terms of how to characterize and describe this pollutant. Technically, atmospheric particulate matter exists in the form of an aerosol. An aerosol is a dispersion of solids and liquids suspended in a gas. In this case the gas is the atmosphere, and the particulate matter is in the form of small droplets of liquids and particles of solids that have many sizes and shapes. Some of the gases in the atmosphere, such as oxides of sulfur and nitrogen, both of which are formed by burning coal, as well as ammonia, also contribute to the formation of particulate matter as described below. Substantial seasonal and geographical variations are also important factors that are relevant to the sources and formation of this pollutant.

Particle size is the most important basis for the classification of particulate matter. Size is important because it affects the exposure and, subsequently, the health effects of particulates. In addition the EPA and other agencies use particle size as the basis for their regulatory and monitoring activities. The concept of the size of particulates would be very simple if they were all perfect spheres and had the same density. Sadly, this is far from the truth. As one might imagine, based on the fact that there are multiple modes for their formation and the many chemicals that are found in them, particulates come in all sizes, shapes, and densities. While some are spherical and easily described in terms of their size, others are elongated fibers, flakes, and virtually any other shape imaginable. To cope with this variability, and in an attempt simplify things a bit, atmospheric scientists use the term *aerodynamic diameter* to describe the size of atmospheric particles. Regardless of their actual size, shape, or density, all particles with the same aerodynamic diameter behave similarly in the atmosphere. More precisely, all particles with the same aerodynamic diameter reach the same final speed as they settle to the ground under the influence of gravity. Even this is somewhat oversimplified. A more detailed definition of aerodynamic diameter is provided in the glossary.

The terms $PM_{2.5}$ and PM_{10} are used to refer to particles with aerodynamic diameters of 2.5 and 10 µm or less. These acronyms are used widely in a variety of publications and reports. Some authors take a shortcut and dispense with the use of the term aerodynamic diameter completely and use diameter, without any other qualification, to refer to particle sizes. Unless otherwise specified, I will use the term diameter to refer to the aerodynamic diameter of particulate matter.

Size and concentration data, expressed in terms of micrograms of a given form of particulate matter per cubic meter of air, are available from the EPA. Epidemiologists and many others have relied heavily on size-based concentration data for their studies of health effects. These data serve as the basis for almost all of the important studies that link particulate matter with long- and short-term health outcomes.

Since particulate matter is an aerosol, most of it enters the body by inhalation. Particle size is a critical factor that determines where inhaled particles are deposited. In general, the larger the particle is, the shorter the distance it travels in the body before it comes to rest after inhalation. Elegant studies of particle deposition have shown that the larger particles (the PM_{10} fraction) are deposited mainly in the nose, nasopharynx, and the larynx [12,13]. The smaller particles, particularly those with aerodynamic diameters of about 2.5 and 0.08 μm, are deposited primarily in the lungs. Particles begin to exert their effects after they enter the lungs [14,15]. In the lung, some particles enter macrophages, sometimes known as the "garbage collectors" that remove contaminants. Others enter cells that are a part of the immune system and thereby begin to activate immunological reactions. Still others enter the blood stream directly, or after they are taken up by macrophages, and thus travel to other organs in the body. These processes are described more fully in chapter 6, which is devoted to the mechanisms by which pollutants affect the body.

There are also data from animal experiments that suggest that up to 20% of the very small particles trapped by the mucous membranes of the nose may enter the brain directly via the olfactory nerve, the nerve that mediates the sense of smell [16]. The end branches of this nerve pass through small holes in the bone, known as the cribiform plate, that separates the nasal cavity from the brain. This creates a direct pathway for small particles to enter the brain from the atmosphere without passing through the lungs.

As a result of these studies, particle size became a medically important criterion for the EPA's particulate matter standards. This is the primary justification for the Agency's current National Ambient Air Quality Standards for particulates that are 10 and 2.5 μm or less in diameter, as shown in table 3.1. Once particles are deposited in the body, the real size of the particle, surface-to-volume ratios, and chemical composition become the factors that affect particle fate and presumably toxicity.

Another method for the classification of particulate matter is based on how it is formed. Primary particulates are formed de novo by combustion. As a rule, these have the smallest diameter when the combustion temperatures are the highest. Primary particulates are also formed by wear and tear of roads, as metals are smelted, from windblown soil, volcanic eruptions, from sea spray, and other sources [17]. Secondary particulates are formed by physical and/or chemical processes acting on components that are already in the atmosphere. The most important precursors are oxides of sulfur and nitrogen, ammonia, and volatile organic compounds [16,18]. Most sulfur dioxide and large portions of atmospheric oxides of nitrogen are the result of burning coal. Ammonia is released into the atmosphere as the result of crop fertilization, particularly with anhydrous ammonia. Some volatile organic compounds in the atmosphere are formed naturally. For example, pinene, the chemical that underlies the aroma of pine forests, is composed of 10 molecules of carbon and 16 of hydrogen [17]. Pinenes are normal components of the resins produced by pine trees and are found in the air near pine forests. Anthropogenic volatile organic compounds come from paints, solvents, and other organic compounds that evaporate and enter the atmosphere [17]. Both naturally occurring and anthropogenic organic compounds are important in the formation of secondary particulate matter. In addition some small primary particles agglutinate or stick together to form larger particles. Droplets of liquids may condense with or be adsorbed onto other preexisting particles.

As one might imagine, the chemical composition of the particulate matter provides another basis for classification. The most common species of particles include sulfates, nitrates, ammonium compounds, black carbon, organic carbon, and soil-derived particles. Black carbon, or soot, is formed during the incomplete burning of fossil fuels, such as coal, oil, and natural gas; forest fires; and other forms of combustion that involve organic compounds. Atmospheric particles also contain a vast array of other compounds and elements.

According to the EPA, approximately 515,000 tons of particles with a diameter of 2.5 μm or less, were emitted by electrical utilities in 2005 [19]. Road dust, miscellaneous processes, and industrial sources were the only sources that exceeded the amount produced by the utilities. The concentration of these smallest particles tends to be the highest in California and in urban areas in the northeastern part of the United States [4].

A nationwide array of monitoring sites records the total amount of the smallest particles detected in a given time period. They do not easily yield data about their source. Some of this information can be gleaned from reports of the chemical composition of particles at various sites.

For example, data gathered by NARSTO (formerly the North American Research Strategy for Tropospheric Ozone, now a broader organization of public and private groups in the United States, Canada, and Mexico) show that the composition of the particles 2.5 µm or less in diameter varies substantially among sites [4]. Almost half of the particles collected at sites in the Ohio Valley and other parts of the eastern United States are in the form of sulfates. This is almost certainly due to the emission of oxides of sulfur produced by burning coal in this region. In this same time period in Toronto, Ontario, where small particle concentrations were similar, only about 20% of these were sulfates. In California, ammonium nitrate is an important component of particulates formed from ammonia released into the atmosphere from the use of fertilizers by large-scale agriculture. The relationships among sulfates, nitrates, and ammonium compounds are complex, due to the dynamic equilibrium that exists in the atmosphere among constituents that form secondary particles [20]. For example, a drop in the concentration of sulfur oxides creates the simultaneous potential for ammonium nitrate concentrations to rise, until ammonia is depleted.

Laden et al. have studied the composition of small particles in six US cities and report finding measurable amounts of silicon, aluminum, calcium, iron, manganese, potassium, lead, bromine, copper, zinc, sulfur, selenium, vanadium, nickel, and chlorine [21]. Using complex statistical techniques, they identified a silicon factor as a marker for particulates arising from the earth's crust, a lead factor for particles from motor vehicle exhaust, and a selenium factor related to burning coal. They then linked the selenium, or coal factor, to daily mortality.

Meteorological conditions also play an important role in determining the concentration of the smallest particles [17]. In the eastern part of the United States, these concentrations are the highest in the summer, whereas in the western part of the country, they tend to be the highest in the winter and fall. There are also changes in the composition of small particles as the result of seasonal changes in fossil fuel combustion and agricultural practices, as well as temperature considerations and the relative concentration of volatile organic compounds in the air from both natural and anthropogenic sources.

It is now possible to measure the concentration of the 2.5 µm particles on a global scale using satellite imaging techniques [22]. Using advanced imaging methods, the global population-weighted geometric mean concentration was found to be 20 $\mu g/m^3$, averaged from 2001 through 2006. To place this value in perspective, the National Ambient Air Quality Standard for the United States is 15 $\mu g/m^3$ (arithmetic average for a year) and the World Health Organization target is 35 $\mu g/m^3$ (annual average). This target value is exceeded over 38% of central Asian populations and 50% of eastern Asian populations. In parts of eastern China, the mean concentration exceeds 80 $\mu g/m^3$. Worldwide and US data in the 2001 to 2006 time interval are shown in figure 3.3.

Mercury

Mercury Sources

Mercury is a naturally occurring heavy metal that is a liquid at room temperature. Mercury has no known useful function in any human metabolic pathway. On the contrary, even small amounts of mercury are toxic, particularly when it is bonded to carbon atoms to form an organic compound, such as methylmercury. Because of its toxicity, mercury has been the subject of intense national and international study and regulations designed to prevent and mitigate its adverse effects, particularly those that affect the developing brain.

The 2008 report issued by the Governing Council of the United Nations Environment Program (UNEP) presents an authoritative description of many of the factors that are necessary to grasp if one is to understand mercury emissions [23]. Since most mercury enters and is disbursed throughout the environment via the air, this book focuses on atmospheric emissions. The UNEP authors identify four interrelated sources or mechanisms by which mercury enters and is spread throughout the environment: primary natural sources, primary anthropogenic sources, secondary anthropogenic sources, and secondary remobilization and re-emission of previously deposited mercury. The UNEP authors estimated that approximately 1,930,000 kg (kilograms) of mercury were discharged into the atmosphere as the result of human activity in 2005 (range 1,230,000–2,890,000 kg). Burning fossil fuels, mainly coal, was the major source of this mercury. This industrial sector alone accounts for about 45% of all anthropogenic emissions, as shown in table 3.3. UNEP authors report that between 400,000 and 1,300,000 kg of dissolved mercury are emitted

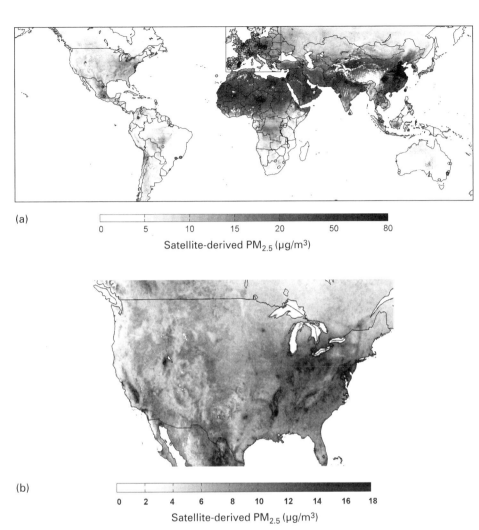

(a)

Satellite-derived PM$_{2.5}$ (μg/m^3)

(b)

Satellite-derived PM$_{2.5}$ (μg/m^3)

Figure 3.3
Satellite-derived concentrations of atmospheric particles with a diameter of 2.5 μm or less averaged between 2001 and 2006. White areas represent over-water sites or locations where there are fewer than 50 observations. (a) Concentrations over the entire earth. Circles identify the location of sites where ground-based comparative measurements were made outside of the United States. The box encloses European sites. (b) Concentrations over the United States and portions of Canada and Mexico. Note the difference in scale between the panels. Maximum concentrations coincide with foci of high populations and coal-fired power plants (see figure 13.2 for sites in United States). Satellite-derived measurements compared well with ground-based measurements at fixed sites ($r = 0.77$, slope = 1.07 for 1,057 comparisons in North America; $r = 0.83$, slope = 0.86 for 244 measurements elsewhere). Aaron van Donkelaar kindly produced these gray-scale images from the pseudo-color images published as figures 3A and 4 in Environ Health Perspect 2010;118:847–55, and these are reproduced with permission from Environmental Health Perspectives and the author [22].

Table 3.3
Worldwide anthropogenic mercury emissions in 2005 in tons (2000 lb)

Region	Combustion from stationary sources, largely coal	Nonferrous metal production	Gold production	Cement production, includes significant coal	Other	Total[a]
Asia, excludes Russia	684	99	65	152	75	1075
Europe, excludes Russia	84	21[b]	0	21	34	160
North America	78	6	14	12	48	158
Russia	51	6	5	4	11	77
Africa	41	2	10	12	3	68
South America	9	15	18	7	6	55
Oceania	21	7	11	1	1	40
World[a]	968	155	122	208	178	1,628

Source: Adapted from table 3.7 of United Nations Environmental Program Technical Background Report to the Global Atmospheric Mercury Assessment [31].

Note: Other sources include pig iron and steel production, primary source mercury production, waste incineration, caustic soda production, and other sources.

a. Apparent summation errors are due to conversion from metric tonnes (2,200 lb) and rounding.

b. Includes pig iron and steel production.

from the oceans and that between 500,000 and 1,000,000 kg are emitted by land sources.

The natural sources of mercury are quite varied. Significant episodic releases occur during volcanic eruptions. This is demonstrated clearly in ice-core samples obtained from the Fremont Glacier in Wyoming, where large peaks in the concentration of mercury were found to coincide with the eruptions of Mount St. Helens and other volcanoes [24]. More constant emissions are due to weathering and erosion of mercury-containing rocks. Geothermal activity also contributes to this global burden of mercury. It is difficult to distinguish the rather constant emissions of mercury from stable natural sources from the mercury that is remobilized and re-emitted. The models of mercury emissions used by UNEP suggest that between one-third and one-half of all mercury emissions are due to these natural sources.

Burning coal, small- and large-scale gold production, cement kilns (which commonly uses coal as an energy source), and other sources are responsible for most of the anthropogenic emissions, as shown in table 3.3.

Secondary anthropogenic releases of mercury occur after it is incorporated into a product or deposited in the environment and then re-emitted as the result of human activity. A partial list of these processes includes incineration of products that contain mercury, such as batteries, paints, switches, and electronics; re-smelting of steel, particularly from automobiles that often contain mercury in their electronics; and chlor-alkalai processes, used to produce chlorine gas and sodium hydroxide. Mercury is also used as a catalyst during the production of vinyl chloride monomer, an important component of plastics. This source is thought to be important yet hard to quantify in China. Small-scale or so-called artisanal gold mining and other dispersed sources are also difficult to quantify. These problems contribute to the uncertainty inherent in estimating mercury emissions. Finally, some mercury is released from dental amalgams when bodies are cremated.

Natural re-emission and remobilization of mercury are difficult to estimate. Once mercury is cleared from the atmosphere and incorporated into soils, and so forth, erosion and fires may lead to a return of previously sequestered mercury into the atmosphere. Since about half of this mercury comes from anthropogenic sources, it is reasonable to conclude that about half of all naturally re-emitted and remobilized mercury is actually attributable to human activity.

China emits more mercury into the atmosphere than any other nation, as shown in table 3.3. China's mercury emissions are thought to be twice as large as the summed emissions from US and Indian sources combined. As might be expected, burning coal to produce electricity accounts for most of this mercury.

Future anthropogenic mercury emissions are difficult to predict as they depend on the degree of economic expansion and success in the control of mercury emissions. The UN mercury report makes three different estimates [23]. In the first, a "status quo" scenario based on the assumption that current emissions will continue unabated, mercury emissions are projected to rise to 1.850 million kg by 2020, an increase of about 25%. Global mercury emissions are projected to fall to about 850,000 kg by 2020 in their "extended emissions control" scenario, in which European standards and promises are extended to all nations. Under a "maximum feasible technological reduction" model, mercury emissions in 2020 could hypothetically be as low as 670,000 kg.

12 - 26 - 15

The Fate of Atmospheric Mercury Emissions

Most atmospheric mercury that comes from natural sources is present in an elemental form (Hg^0). Mercury released by combustion is typically present in one of three different forms: elemental mercury, a divalent reactive gaseous form (Hg^{++}), and mercury that is bound to particles [25]. The lifetime of elemental mercury in air is thought to be between six and eighteen months. This is the form that is most likely to travel long distances before deposition. Other forms, especially those bound to particles, have shorter lifetimes in air and appear to travel shorter distances before deposition.

In 2002 the EPA funded a monitoring site in Steubenville, Ohio, that was designed to determine the source(s) of the mercury deposited at that location [25]. The site was chosen because there are 17 major anthropogenic sources of mercury within 100 km of the monitoring equipment. State-of-the-art data collection and statistical modeling techniques were used in conjunction with meteorological data to differentiate among potential sources of mercury. These investigators concluded that approximately 70% of the mercury deposited at the site was derived from

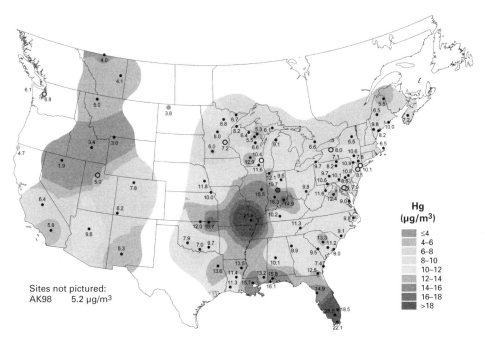

Figure 3.4
US mercury deposition rates in 2008. Available from the National Atmospheric Deposition Program, Mercury Deposition Program [26]. Note: This is a black and white version of the full-color original.

burning coal at local and regional sites [25]. The National Atmospheric Deposition Program, a cooperative effort by various governmental, educational, and other agencies, reports substantial differences in mercury deposition in the United States, as shown in figure 3.4 [26]. Monitoring sites in south Florida and along the Mississippi and Ohio rivers report the largest depositions (up to 25 μg/m^2 in south Florida).

Atmospheric mercury returns to the earth by several mechanisms. Some is incorporated into precipitation, some falls as dust, and some descends due to the action of gravity. The impact of this deposited mercury depends in part on its form. Ionic mercury (Hg^{++}) has a higher impact than elemental mercury (Hg^0). Once deposited, mercury is methylated by the action of bacteria. This conversion is favored in water that is acidic and contains large amounts of dissolved organic material, such as that found in many swamps and waterways along the Atlantic coastal plain [27]. Methylmercury is highly diffusible and enters the food chain via microscopic plants and animals. There, this persistent toxin bioaccumulates and is biomagnified. This effect is shown clearly by Driscoll et al., who measured mercury and the fraction that is present as methylmercury in the food chain [28]. Both increase as mercury moves up the food chain, with the lowest concentration and the lowest fraction as methylmercury found in water. These increase progressively in phytoplankton < zooplankton < plankton-eating fish < fish-eating fish < loons. Loons eat fish that are usually ten inches in length or less and have methylmercury concentrations that are as much as one million times those found in water. Luckily for us and the loons, we do not eat them! We do eat fish. In Wisconsin, the methylmercury concentration in loon eggs is greater than the concentration that is toxic to laboratory animals.

The process of mercury deposition, conversion to methylmercury, and biomagnification has led state and tribal authorities to issue fish consumption advisories in every state [29]. Minnesota leads the list with 1,203 advisories, while the District of Columbia has only one. These advisories include 43% of all of the nation's total lake acres, excluding the Great Lakes, 39% of the nation's river miles, and 42% of the coastal waters in the lower 48 states, increasing to 79% if Alaska and Hawaii are included.

Methylmercury concentrations are therefore the highest in large predatory fish and marine mammals, where concentrations may be several orders of magnitude greater than in the surrounding water. This principle is also shown in the measured concentration of mercury-in-fish maps published by the EPA [30]. The highest concentrations were measured in the lower Mississippi River, south Florida, and the Atlantic coastal plain in the Carolinas.

Information concerning the concentration of mercury in fish is available from the Food and Drug Administration [29]. Large tuna, tilefish from the Gulf of Mexico, swordfish, and king mackerel typically contain the highest concentrations of mercury. Fish and shellfish low in mercury include tilapia, freshwater trout, scallops, sardines, and pollock.

References

1. Study of Hazardous Air Pollutant Emissions from Electric Utility Steam Generating Units—Final report to Congress EPA publication 453/R-98–004a. Washington DC: EPA, 1998.

2. US Environmental Protection Agency. History of the Clean Air Act, Available at http://www.epa.gov/air/caa/caa_history.html. 2010.

3. Agency for Toxic Substances and Disease Registry (ATSDR). ToxGuide for Arsenic. Atlanta: Department of Health and Human Services, 2010.

4. Blanchard C. Spatial and temporal characterization of particulate matter. In: McMurry PH, Shepherd MF, Vickery JF, eds. Particulate Matter Science for Policy Makers: a NARSTO Assessment. Cambridge: Cambridge University Press, 2004.

5. State of California Air Resources Board. Proposed Identification of Inorganic Arsenic as a Toxic Air Contaminant. Sacramento, 1990.

6. Integrated Science Assessment for Sulfur Oxides—Health Criteria. National Center for Environmental Assessment. Washington DC: EPA, 2008.

7. National Research Council. Hidden Costs of Energy: Unpriced Consequences of Energy Production and Use. Washington DC: National Academy of Sciences, 2009.

8. Rodhe H, Dentener F, Schulz M. The global distribution of acidifying wet deposition. Environ Sci Technol 2002;36(20):4382–8.

9. Boiler Emissions Reference Guide, 2nd ed. Milwaukee: Cleaver Brooks, 2010.

10. Integrated Science Assessment for Oxides of Nitrogen—Health Criteria. Research Triangle Park NC: US Environmental Protection Agency, National Center for Environmental Assessment, 2008EPA/600/R-08/071. Washington DC: EPA, 2008.

11. Allen J. Chemistry in the sunlight—a NASA report. NASA Goddard Spaceflight Center, Greenbelt, MD, 2002. Available at http://earthobservatory.nasa.gov/Features/ChemistrySunlight/.

12. World Health Organization. World Health Report 2002. Geneva: WHO, 2002.

13. McClellan R, Jessiman B. Health context for management of particulate matter. In: McMurry PH, Shepherd MF, Vickery JF, eds. Particulate Matter Science for Policy Makers: a NARSTO Assessment. Cambridge: Cambridge University Press, 2004.

14. Brook RD, Rajagopalan S, Pope CA, III, et al. Particulate matter air pollution and cardiovascular disease: an update to the scientific statement from the American Heart Association. Circulation 2010;121(21):2331–78.

15. Peters A, Veronesi B, Calderon-Garciduenas L, et al. Translocation and potential neurological effects of fine and ultrafine particles a critical update. Part Fiber Toxicol 2006;3:13.

16. Oberdorster G, Sharp Z, Atudorei V, et al. Translocation of inhaled ultrafine particles to the brain. Inhal Toxicol 2004;16(6–7):437–45.

17. Integrated Science Assessment for Particulate Matter: First External Review Draft. Research Triangle Park NC: National Center for Environmental Assessment, 2007.

18. Hidy G, Niemi D, Pace T. Emission characterization. In: McMurry PH, Shepherd MF, Vickery JF, eds. Particulate Matter Science for Policy Makers: A NARSTO Assessment. Cambridge: Cambridge University Press, 2004.

19. US Environmental Protection Agency. Air Emission Sources: Particulates. Washington DC: EPA, 2011.

20. Pandis S. Atmospheric aerosol processes. In: McMurry PH, Shepherd MF, Vickery JF, eds. Particulate Matter Science for Policy Makers: A NARSTO Assessment. Cambridge: Cambridge University Press, 2004.

21. Laden F, Neas LM, Dockery DW, Schwartz J. Association of fine particulate matter from different sources with daily mortality in six U.S. cities. Environ Health Perspect 2000;108(10):941–7.

22. van Donkelaar A., Martin RV, Brauer M, et al. Global estimates of ambient fine particulate matter concentrations from satellite-based aerosol optical depth: development and application. Environ Health Perspect 2010;118(6):847–55.

23. UNEP Chemicals Branch. The Global Atmospheric Mercury Assessment: Sources, Emissions and Transport. Geneva: UNEP-Chemicals, 2008.

24. Schuster PF, Krabbenhoft DP, Naftz DL, et al. Atmospheric mercury deposition during the last 270 years: a glacial ice core record of natural and anthropogenic sources. Environ Sci Technol 2002;36(11):2303–10.

25. Keeler GJ, Landis MS, Norris GA, Christianson EM, Dvonch JT. Sources of mercury wet deposition in eastern Ohio, USA. Environ Sci Technol 2006;40(19): 5874–81.

26. National Atmospheric Deposition Program. Available at http://nadp.sws. uiuc.edu. Accessed 2010 May 5.

27. Hughes WB, Abrahamson TA, Maluk TL, Reuber EJ, Wilhelm LJ. Water quality in the Santee River Basin and coastal drainages, North and South Carolina, 1995–98: US Geologic Survey Circular 1206. USGS, 2000.

28. Driscoll CT, Han Y-JCCY, Evers DC, et al. Mercury contamination in forest and freshwater ecosystems in the northeastern United States. BioScience 207;57(1):17–28.

29. US Food and Drug Administration. Seafood, 2011. Available at: http:// www.fda.gov/Food/FoodSafety/Product-SpecificInformation/Seafood/ default.htm.

30. EPA Mercury Maps: A Quantitative Spatial Link between Air Deposition and Fish Tissue. Peer-Reviewed Final Report. Washington DC: EPA, 2001.

31. AMAP/UNEP. Technical Background Report to the Global Atmospheric Mercury Assessment. Geneva: UNEP-Chemicals Branch, 2008.

4

From Mine to Ash

The era of coal-fired power plants in the United States began at 3:00 pm on September 4, 1882, when Thomas Edison threw a switch that turned on the lights in J. Pierpont Morgan's office. This began the delivery of electricity from the Pearl Street Station, located about a half mile away. This generating station operated until 1890, when, ironically, it burned down. Soon huge amounts of coal were needed to fuel the boilers and large amounts of ash were produced. This is how the modern life cycle of coal began: from mine to ash. Although most discussions of coal and health effects are focused appropriately on the smokestack pollutants, a complete discussion must include the health implications of the mine-to-ash pathway.

To illustrate the mine–ash pathway, I will use the Tennessee Valley Authority Kingston Fossil Plant in Tennessee as a typical example of an aging coal plant. The plant gained notoriety in 2008 because of the failure of a dam that retained coal ash slurry. The Kingston Plant was built to supply electricity to the nearby Oak Ridge National Laboratory. Like virtually all other coal plants, large amounts of water are needed to cool the steam in the condenser. Therefore the designers located the plant at the intersection of the Clinch and Emory rivers near Kingston.

Construction was started in 1951. When it was completed in 1955, it was the largest coal fired electricity generating facility in the world, a distinction it held for ten years [1]. The plant houses nine boilers: four are rated at 200 megawatts and five are rated at 175 megawatts [2]. The original nine relatively short smoke stacks are still present but have been replaced by two that are much taller. Reports supplied to the EPA indicate that the total output of the Kingston Plant was just over 10.9 million megawatts in 2008 [3]. When it operates at full capacity, it burns about 14,000 tons of coal each day [1].

Burning that much coal produces enormous amounts of waste. Reports submitted to the EPA document the release of 11.07 million tons of carbon dioxide in 2008, an amount that is similar to emissions for the years between 1996 and 2008 [1,3]. As a result of emission controls required by the Clean Air Act and the Acid Rain Program, 2008 nitrogen oxide emissions of 7,928 thousand tons were about 22% of the peak value of 36,240 tons emitted in 1976, and 2008 emissions of sulfur oxides of 50,617 tons were about 27% of the peak value of 191,000 tons emitted in 1975 [1,3]. Large amounts of other pollutants are discharged into the atmosphere and into the Clinch and Emory rivers, as shown in table 4.1. These releases do not appear

Table 4.1
Emissions from TVA Kingston Fossil Plant, 2008 data

Pollutant	Emissions from stacks	Discharges to Clinch River	Discharges to Emory River
CO_2	11.07 million tons		
$PM_{2.5}$	3,056.6 tons[a]		
SO_2	50,600 tons		
NO_x	7,930 tons		
Hydrochloric acid	1,550 tons		
Hydrofluoric acid	225 tons		
Arsenic compounds	250 lbs	2,600 lbs	70 tons
Barium compounds	2,900 lb	21 tons	
Chromium compounds	250 lbs		3.15 tons
Cobalt compounds	250 lbs		23 tons
Copper compounds	250 lbs	4,000 lbs	46 tons
Lead compounds			35.5 tons
Manganese compounds	750 lb		375 tons
Mercury compounds	480 lb		400 lbs
Nickle compounds	250 lbs	250 lbs	36.5 tons
Selenium (2007 data)		1,100 lbs	
Vanadium compounds	250 lbs	3,200 lbs	320 tons
Zinc compounds	1,500 lbs	4,000 lbs	90 tons

a. 2005 data, 2008 data not available, 2005 National emissions inventory data and documentation, available at http://iaspub.epa.gov/enviro/tris_control .tris_print?tris_id=37763STVKNSWANP.

to be attributable to the 2008 coal ash spill as similar amounts were released in prior years. A perusal of reports required of the Kingston Plant reveals some surprises—for example, in 1999 the facility reported transferring coal ash that contained 1,600 lbs of barium compounds, and 250 lbs each of lead, arsenic, chromium, cobalt, copper, and manganese compounds to the Oakdale School in Oakdale, TN. When contacted, school officials seemed unaware of this transfer (personal communication).

Mining Coal

Mining, and underground mining in particular, has always been a dangerous occupation. Mining coal is no exception. Mine explosions and other mining disasters, particularly those that occur in North and South America, Europe, and other places, such as New Zealand, invariably make headline news. Mining disasters are spectacles. An underground location adds to the psychological drama. Mining disasters in China are much more common, but rarely attract the attention of the Western media.

In the United States, mining coal is regulated by The Surface Mining Control and Reclamation Act (30 USC §§1234–1328). This is the major statute that governs coal mining operations. It was adopted in 1977 to provide federal control of mining operations. This Act established environmental standards for surface mines, required permits and land reclamation bonds, and authorized governmental regulators to inspect mines and punish violators.

The term "mining accidents" is usually inappropriate. These are human-made disasters, not accidents. Too frequently they are the result of a disregard for safety practices, a violation of regulations, or a combination of these factors. The constant push to increase profits is and always has been powerful.

Coal mines are divided broadly into two types—surface and underground mines. The choice of a mining technique depends on a number of factors that ultimately relate to the cost of the operation and the value of the mined coal, that is, profitability. Health effects associated with mining coal relate directly to the impacts on miners themselves and the impact on society as a whole due to the environmental consequences of coal mining, transport, washing, burning, and, finally, the disposal of the residual ash.

❧   Surface mines have a number of common features. Generally, the coal is relatively close to the surface and is mined after removal of the soil and rock. In strip mines, the rock, or overburden, that covers the bed of coal must be removed to an adjacent site. Huge pieces of earth-moving equipment are built specially for this purpose. Then the exposed seam of coal can be mined. The process is repeated until the coal seam is exhausted or it is too thin, deep, or otherwise unprofitable to mine. Open-pit mines are similar to strip mines in that the overburden is removed to an adjacent site to expose the coal, thereby creating an enormous hole in the ground, or pit. The coal seams in open-pit mines, such as those in the Powder River Basin, located in southeast Montana and northeast Wyoming, may be very thick, making it difficult, if not impossible, to reclaim the land. Both strip and open-pit miners may make extensive use of blasting to remove the overburden or loosen the coal.

❧   Mountain-top removal valley-fill mining is another form of surface mining. The top of the mountain, or overburden, is removed and dumped into an adjacent valley. The exposed coal is then removed, leaving behind a plateau and a filled valley. This form of mining is increasingly common in West Virginia and Kentucky and very controversial because valleys and their ecosystems are destroyed and water supplies may be damaged.

❧   Underground mining techniques are used to gain access to seams of coal that are buried far below the surface. So-called long-wall mining accounts for about half of all underground coal mining. Once the shaft reaches the seam of coal, a variety of strategies are used to loosen and remove the coal. In one process, a rotating drum equipped with teeth moves back and forth along the length of the seam, loosening the coal from the seam and dumping it onto a conveyor belt. The mined coal is then removed from the mine. Temporary roof supports are used to support the ceiling of the mine. Controlled collapse of the mine may be allowed to occur as the machine advances. Typically this is a highly mechanized, semi-robotic process that is said to be among the safest of the underground methods. Other somewhat related methods for supporting the roof of the mine may be used.

All mines pose problems associated with ecosystem damage. Aside from the obvious damage caused by large-scale excavations and disposal of overburden, mining exposes large surfaces of rock that have never been subjected to weathering. This creates the potential for leaching dissolved minerals and heavy metals into the water supply and the disruption of the acid–base balance of adjacent water systems.

Large amounts of methane are released when coal is mined. Methane, along with coal dust, is responsible for explosions in coal mines and the asphyxiation of miners or others who enter mines where methane has displaced air. Mining coal is the fourth-ranked anthropogenic source of methane discharges into the atmosphere (Methane Sources and Emissions, available at htpp://www.epa.gov/methane/sources.html). Methane makes an important contribution to global warming. In some modern mines, methane may be trapped and either burned or otherwise used to improve the efficiency of mine operation.

The environmental consequences of mining have been studied intensely in the portions of West Virginia, Virginia, and North Carolina that are included in the Kanawha and New River basins [4]. Not surprisingly, there were large disruptions of the animal and plant life in the streams in this region. The concentration of *E. coli* also exceed national guidelines in 26% to 43% of samples. *E. coli* is a bacterium that may cause serious or fatal gastrointestinal diseases. In addition there were also significant increases in dissolved sulfate concentrations that were highest in the counties where mining was most intense, measurable concentrations of volatile organic compounds in the water, and increases in the concentrations of nickel, chromium, and zinc that were high enough to harm aquatic life. The groundwater concentrations of radon were among the highest in the United States. Finally, methane concentrations reached potentially explosive levels in 7% of the drinking water wells in the region.

Although there have been dramatic improvements in coal mine safety, it is still a dangerous occupation. Despite the explosion April 5, 2010, in the Upper Big Branch Mine, operated by Massey Energy, that killed 29 miners, mining coal is much safer in the United States than in other countries, particularly China, where official sources reported almost 6,000 deaths in 2005 [5]. According to the US Bureau of Labor statistics, there were 24.8 deaths per 100,000 full-time equivalent coal miners compared to 4.3 deaths per 100,000 full-time employees in all of private industry [6].

"Black lung disease," or more properly coal workers' pneumoconiosis, is undoubtedly the most infamous of the occupational diseases affecting coal miners. Coal workers' pneumoconiosis is a chronic lung disease caused by the inhalation of particles created by mining coal. The most severe form is called progressive massive fibrosis. As one might expect, the risk for the development of this lung disease increases the longer one works in a mine. Among miners who are 60 and older, the prevalence of coal workers' pneumoconiosis is just over 5% and the prevalence of the more severe fibrotic form is about 0.8% [7]. Both forms are more prevalent among underground miners than among surface miners. For both groups of workers, these diseases are more common among those who mine anthracite than lower ranked coals.

The incidence of both the fibrotic and nonfibrotic forms of black lung disease decreased steadily between 1987 and 2002 as a result of regulations imposed by the Federal Coal Mine Health and Safety Act of 1969 [7]. Unfortunately, that trend has not held true. The most recent data show that the prevalence of X-ray evidence of coal workers' pneumoconiosis has increased in miners in mines of all sizes [8]. The problem is the most severe among miners who work in small mines (fewer than 50 employees) where the rates are five times higher than among miners who work in large mines [8]. According to Seaton, this increase is attributed to a relaxation of regulatory activity due to lobbying efforts of "the wealthy who control the media" [9].

For quite some time it was hypothesized that the amount of dust inhaled and possibly the amount of silicon dioxide (SiO_2, a major component of sand and quartz) determined the risk for the development of coal workers' pneumoconiosis. More recently, this hypothesis has been revised and the amount of bioavailable iron in coal dust appears to be the factor that determines the likelihood of developing coal workers' pneumoconiosis [10,11]. For a more complete discussion of this disease, see chapter 8 on respiratory diseases.

Washing Coal

After coal is mined, it is usually washed or otherwise purified before it is shipped to its final destination. This step separates the coal from dirt and rock and may remove substantial amounts of sulfur. Typically washing processes rely on the physical differences between coal and the undesired components. Many of these are based on differences in the weight of coal per unit of volume compared to the weight per unit of volume of the impurities. Large centrifuges, fluid media with different densities, and other physical techniques are employed. Washing makes the product cheaper to ship and cleaner to burn and is one of several elements of what some refer to as "Clean Coal." Disposal of the resulting waste or slurry, however, is a substantial challenge.

Just before 8:00 am on February 26, 1972, a heavy equipment operator working at the Buffalo Creek Mine in West Virginia noticed that the highest of three dams that had been built to contain coal-wash slurry was "real soggy" [12]. Five minutes later, the dam collapsed. The resulting floodwaters rushed down Buffalo Creek Hollow, inundating the two lower dams, releasing an estimated 132 million gallons of the black watery waste. Within minutes, 125 people were dead, 1,100 were injured, and 4,000 were homeless. This was one of the worst flooding disasters in US history.

This was not an accident, nor was it an "Act of God" as claimed by coal company officials who said that the dam "was incapable of holding the water that God poured into it." The three investigations that followed the disaster showed that the mine operators had "blatantly disregarded standard safety practices" [12]. The state of West Virginia sued mine operators for $100 million. Arch Moore, West Virginia's governor at the time, negotiated a $1 million settlement three days before he left office. The lawsuits that ensued netted each individual about $13,000. The lawyers for the plaintiffs donated a portion of their fees to construct a community center. It was never built.

On October 11, 2000, near Inez, Kentucky, the bottom of a 72-acre coal waste impoundment at a mountaintop coal mine collapsed. Millions of gallons of slurry were released into an abandoned mine from which it entered and fouled about 60 miles of the Big Sandy River [13]. The effects of this breech are still present more than ten years later [14]. This facility was operated by Martin County Coal Corp., a subsidiary of Massey Energy, the same company that operated the mine in West Virginia where 29 miners died in 2010.

These are not isolated incidents. According to the Coal Impoundment Location and Information System, maintained by Wheeling Jesuit

University, there have been 65 impoundment failures that released over 748 million gallons of slurry since the 1972 Buffalo Creek disaster [15]. As a result of the Inez release, Congress charged the National Academy of Science to "examine engineering practices and standards currently being applied to coal waste impoundments and to consider options for evaluating, improving, and monitoring the barriers that retain coal waste impoundments" [16]. The Committee made numerous recommendations, including, "...that the total system of mining, preparation, transportation, and utilization of coal and the associated environmental and economic issues be studied in a comprehensive manner to identify the appropriate technologies for each component that will eliminate or reduce the need for slurry impoundments while optimizing the performance objectives of the system" [16]. This is a substantial task that has yet to be completed.

Most coal slurry is stored using one of two methods: impoundment behind dams, the most common strategy, or injection into abandoned mines. Regardless of the storage method, there is concern that chemicals will leach from the coal into the groundwater where they have the potential to affect health. Large-scale studies are lacking, but many are concerned that these practices are not safe [17].

One approach to understanding the possible health effects of slurry created by washing coal has been taken by a team of investigators from the US Geological Survey [18]. This team began with the observation that kidney cancers were common in Louisiana. This, combined with the hypothesis that a chronic kidney disease in which cancer is common (Balkan Endemic Nephropathy) is more prevalent in regions where lignite is found, led to tests of wells in the part of Louisiana where lignite is mined [19]. Lignite is a low-ranked coal that contains potential carcinogens in the form of volatile organic compounds and heavy metals (see chapter 2). In this study there was a significant association between a number of coal-associated elements and chemicals found in well water and cancers of the kidney. These included organic components, phosphate, ammonia, arsenic, boron, bromine, chlorine, chromium, fluorine, lithium, sodium, phosphorous, rubidium, selenium, strontium, and tungsten [18]. The fact that coal slurry contains many of these elements and compounds provides additional support for the hypothesis that water contaminated by chemicals in coal slurry poses a significant risk to human health.

Invariably, the washing or preparation of the coal leads to the loss of coal itself. Increasing the efficiency of these processes is one of the major challenges in this industry. Fluidized bed combustion and gasification is one of the processes designed to minimize this waste. In these boilers the "fuel" is suspended by jets of air containing controlled amounts of oxygen and steam to produce synthesis gas or syngas. The syngas is then burned with the overall expectation that total emissions and coal wastage are reduced.

Transporting Coal

Although many of the newer coal-fired power plants are built adjacent to the mines that supply them, huge amounts of coal are transported from mines to generating stations. While over 70% of coal tonnage is moved by rail, significant amounts are moved by truck, via waterways, or by some combination of truck, rail, and waterway [20]. Coal transportation statistics are shown in table 4.2. The health impacts of coal transportation are related primarily to emissions from diesel locomotives and railroad accidents.

Most diesel locomotives are so-called diesel-electric locomotives. They use a diesel engine to generate electricity that is then used to power electric motors that move the train. This allows the diesel engine to run at near-constant speeds and at peak efficiency. Locomotives are of several types: line-haul, passenger, and switch locomotives. Switch locomotives

Table 4.2
Coal shipment characteristics [20]

Transport mode	Value ($ millions, % of total)	Tons (thousands, % of total)	Ton-miles (millions, % of total)	Average miles per shipment
Rail	25,907, 67.8%	1,015,735, 71.7%	773,290, 92.5%	428
Truck	7,179, 18.8%	217,602, 15.4%	14,002, 1.7%	61
Water	1,701, 4.4	43,925, 3.1%	6,548, 0.8%	145
Multiple modes	1,832, 4.8%	52,352, 3.7%	39,389, 4.7	340

often spend substantial amounts of time with the engine idling, thus emitting pollutants while doing no work. Hybrid locomotives, known as "Green Goat" and "Green Kid" switch engines, have been introduced to reduce emissions and improve efficiency.

Diesel engines are important sources of particulate matter and oxides of nitrogen. The EPA recognized their threats to health and finalized new emission standards for locomotives and marine diesels in March 2008 [21]. This followed on the heels of a rule adopted in May 2004 that reduced the sulfur content of the fuel used for non-road diesel engines by 99% (Code of Federal Regulations, parts 9, 69, et al. OAR-2003–012). The EPA estimated that the locomotive rule would reduce particulate matter emissions by about 90% and nitrogen oxide emissions by 80%, compared to the standards in place at the time the rule was adopted. A 0.6% rail transportation cost impact was predicted by the Agency. The monetized health benefits by 2030 were estimated to be $9.2 to $11 billion, yielding a benefit-to-cost ratio between 9:1 and 15:1, depending on the method used to make the calculation. These dollar figures are important and powerful indicators of the negative health impacts of the pollutants from this often-ignored source.

According to the US Department of Transportation, there were 8,427 train accidents or incidents between January and September 2010 [22]. Fatalities occurred in 557 of these (6.61%). The frequency of train accidents was 2.67 for each million train-miles with 1.13 deaths per million train-miles. Human factors and track defects each accounted for about a third of these accidents with equipment defects and other causes about equally split among the remainder. The deaths per mile for trains compares unfavorably with automobile statistics. The National Highway Transportation Safety Administration reported 74 injuries with 1.13 deaths per 100 million vehicle miles traveled in 2009 [23].

Burning Coal

The coal plant is where the rubber meets the road—the place where coal is burned, transforming a mineral into electricity and a myriad of pollutants. Conceptually these are simple units. Coal is burned and the heat is used to generate steam that powers a turbine. The turbine rotates a generator that produces electricity.

In reality, coal plants are huge, highly engineered, complex, and very expensive facilities. Goodell compares the visual experience of entering the combustion chamber of a boiler undergoing maintenance to that of entering the nave of a cathedral—both are vast with prominent vertical elements [24]. The coal is typically delivered to the site by rail and stored temporarily. From the storage site, the coal is taken to a mill where it is pulverized into a fine powder. The powder is sprayed into a huge combustion chamber where it creates an intensely hot fireball (vivid videos of this are posted on YouTube). Most coal plants use powdered coal because of the relative ease with which combustion can be controlled.

Some newer plants use fluidized bed combustion technology. In these boilers the fuel is suspended by jets of air, along with additives such as limestone, that trap sulfur in the resulting ash. These boilers are easier to control and operate at lower temperatures, a condition that minimizes the formation of NO_x [25]. This is another element of so-called clean coal technology.

The heat from the burning coal vaporizes water in tubes in the boiler, converting it to steam at very high pressures. The superheated steam rotates a turbine connected to a generator to produce the electricity. After the steam leaves the turbine, it is still very hot and must be cooled in huge condensers. These condensers require large amounts of cooling water. This is why many coal plants are built near rivers or other large bodies of water. The cooled condensate is returned to the boiler and the cycle continues.

The flue gases from the boiler pass through pollution control devices (described in chapter 5). The partially purified flue gas is emitted into the air via stacks that are usually very tall to dilute and disperse the gas stream.

Disposing of Waste

Burning coal produces huge amounts of waste. As shown in table 2.1, between 6 and 10% of coal, by weight, is ash. This is what is left behind after coal is burned and is known variously as coal ash, coal combustion waste, or coal combustion residue. In addition to the ash that is an inherent part of coal, control devices that remove pollutants, such as sulfur-containing compounds, produce solids that contribute to this waste stream. Pollution mitigation devices are described in some detail in chapter 5. US power plants produce at least 120 million tons of ash each year [16]. Based on estimates of worldwide coal consumption, about 520 million tons of ash are produced each year, enough to fill over 4 million railroad cars [26,27]. Coal ash is thus one of the major waste streams in industrialized societies.

Few Americans knew anything about coal ash before December 22, 2008. Everything changed that day. At about 1:00 am that morning, a dike at a coal ash settling pond adjacent to the Kingston Fossil Plant, located near Kingston, Tennessee, ruptured, releasing about 5.4 million cubic yards of coal ash slurry into the Emory and Clinch rivers and the surrounding area. According to congressional testimony by an EPA official, the spill inundated about 300 acres, destroyed three homes, disrupted electrical power in the area, and severed a natural gas pipeline [28]. In addition the spill covered roads and railroad tracks. Immediately after the spill, the EPA, the Tennessee Valley Authority, the operator of the Kingston Fossil Plant, and the Tennessee Department of Environmental Conservation began to sample air, surface water, and drinking water to determine whether there were threats to health. The EPA found levels of arsenic, cobalt, iron, and thallium in residential soil samples that were above residential Superfund screening levels and EPA's Residential Removal Action Levels [28]. River water samples showed elevated levels of arsenic, cadmium, chromium and lead immediately after the incident and again after a rainstorm in January of 2009 [28]. The concentrations of heavy metals were below drinking water limits in follow-up samples. There was a considerable amount of controversy concerning these results. Appalachian Voices and Watauga Riverkeeper collected samples that contained arsenic, barium, cadmium, chromium, lead, mercury, nickel, and thallium in concentrations that exceeded one or more Tennessee water quality standards [29].

Coal ash is generally divided into two types, based on where it forms and is collected. Bottom ash forms in the combustion chamber of the power plant and drops, by gravity, to the bottom of the boiler where it is quenched in water. The ash, clinker, or slag, as it may be called, is ground into smaller pieces and removed continuously by one of several mechanical mechanisms. The water-ash mixture is moved to a settling pond where it is impounded. Fly ash is composed of much smaller particles. These are suspended in the boiler's exhaust gases. The particles created by flue gas desulfurization procedures also become a part of the fly ash. Fly ash particles are spherical and contain large amounts of silicon dioxide (SiO_2) and calcium oxide (CaO). Fly ash also contains significant amounts of other metals that are usually present in concentrations that are higher than those found in bottom ash. Median values for fly ash constituents include the following (in parts per million): arsenic, 43.4; barium, 807; beryllium, 5.0; boron, 311; cadmium, 3.4; chromium compounds (including chromium VI), 226; cobalt, 35.9; copper, 112; lead, 56.8; manganese, 250; mercury, 0.1; nickel, 77.6; selenium, 7.7; strontium, 775; thallium 9.0; vanadium, 252; and zinc, 148 [26].

Coal also contains measurable amounts of uranium, thorium, and radon, a decay product of uranium. Concentrations are typically ten times those found in the coal before combustion [30]. Radon is a gas, and exits the power plant in flue gases. The much heavier uranium and thorium, along with their decay products, are found in the fly ash where uranium concentrations are typically between 10 to 30 ppm, a concentration often found in many naturally occurring shales and granites [30]. The US Geological Survey concludes that these concentrations account for less that 1% of all radiation exposures to the average American [30].

In the United States most coal ash is disposed of in landfills or surface impoundments at the site of the power plant. This is the most economical way to deal with it because it is expensive to move it off site. In its coal ash risk assessment, the EPA describes three types of coal combustion waste [31]. Conventional waste includes fly ash, bottom ash, boiler slag, and flue-gas desulfurization sludge. This is the most common type of waste and is produced by plants that use powdered coal. The second type is called codisposed coal combustion waste and coal refuse. This is a mixture of ash and the slurry from washing coal. The third type is fluidized bed combustion waste. This waste is quite alkaline due to the addition of limestone in the combustion chamber. These wastes are typically disposed of in one of three different kinds of sites: those that are unlined, those lined with clay, or those lined with composite materials. Most waste disposal sites constructed after 1995 are lined. Those constructed before then are a mixture of lined and unlined. Other disposal methods include off-site landfills and impoundments or returning the ash to a nearby mine. Additional ash may be "recycled" by incorporating it into wallboard or cement. Because of its chemical properties, coal ash can replace substantial amounts of Portland cement in concrete. Coal ash produced in coal-burning cement kilns may be incorporated directly into the cement without ever becoming a part of a waste stream.

As a part of its obligation to protect human health and the environment and growing concerns about the hazards posed by coal ash disposal practices, the EPA has undertaken studies of coal ash as a prelude to rulemaking. As a result of this process, two important documents were released in 2007. The first of these was a damage assessment [32]. In this report the EPA detailed the results of their evaluation of 85 potential damage cases. Twenty-four were proven cases of damage. Typically the damage was in the form of contamination of surface or groundwater supplies. These were due to disposal in unlined sand and gravel pits, unlined landfills, unlined surface impoundments, or due to a liner that failed at a surface impoundment. An additional 43 cases were found to be locations where there was potential but not proven damage to surface or groundwater.

The findings in the Town of Pines in Porter County, Indiana, are representative of a proven damage case [32]. At this site about 1 million tons of fly ash were buried in a landfill and used as fill at various construction sites. In 2000 a resident in that area filed a complaint with the Indiana Department of Environmental Management. Water from a private well had a foul taste. Subsequent testing found that residential wells were contaminated by elevated levels of a number of toxins, including arsenic, manganese, and volatile organic compounds. The Department concluded that leaching of heavy metals, including manganese, boron, molybdenum, arsenic, and lead from the coal ash landfill was responsible for some of the contaminants. The volatile organic compounds were attributed to other sources. The matter was resolved via an Administrative Order of Consent that brought municipal water to 130 homes with private wells and required additional remediation of the site.

The second of the reports on coal ash issued by the EPA in 2007 was a health and ecological risk assessment [31]. The conclusions presented in this lengthy complex document are as follows: risks to health are greatest from constituents stored in unlined surface impoundments, lower from the impoundments lined with clay, and the lowest from those lined with composite materials. Lining surface and landfills with composites reduced risks for cancer and noncancer effects to below threshold criteria. Most post-1995 disposal sites are lined with composites; however, they may fail at some time in the future. No liner lasts forever. The greatest risk was posed by arsenic from unlined surface impoundments. In the screening analysis the most highly exposed (90th percentile) had a cancer risk of 9×10^{-3} for those exposed by the groundwater to drinking water pathway (where the cancer criterion was an excess risk that was greater than one chance in 100,000 or 10^{-5}). In general, risks were lower in the full-scale analysis as shown in table 4.3.

Table 4.3
Full-Scale coal combustion waste human risk results at the 90th percentile of exposure, groundwater to drinking water pathway [31]

	90th Percentile hazard quotient or cancer risk value	
Constituent	Unlined units	Clay-lined units
Landfills		
Arsenic (cancer risk)	5×10^{-4}	2×10^{-4}
Thallium HQ	3	1
Antimony HQ	1	0.6
Molybdenum HQ	1	0.7
Lead (ratio of exposure to MCL)	0.9	0.2
Cadmium	0.3	0.3
Boron	0.5	0.3
Selenium	0.4	0.2
Nitrate/nitrite (ratio of exposure to MCL)	0.2	0.2
Surface impoundments		
Arsenic (cancer risk)	9×10^{-3}	3×10^{-3}
Molybdenum	5	3
Cobalt	5	0.9
Cadmium	5	1
Lead (ratio of exposure to MCL)	5	0.9
Boron	3	2
Selenium	1	0.8
Nitrate/nitrite (ratio of exposure to MCL)	0.2	0.7

Note: MCL = maximum contaminant level, the legal threshold limit on the amount of a substance that is allowed in public water systems under the Safe Drinking Water Act. HQ = hazard quotient, the ratio of the potential exposure to the substance and the level at which no adverse effects are possible. For an HQ greater than one, a risk is possible but not certain—can't be translated to a risk probability. For an HQ less than one, no risk is expected.

In the aftermath of the Kingston coal ash spill and other concerns about the integrity of systems that retain coal ash, the EPA sent letters to facilities in order to better evaluate the safety of current storage practices [33] The agency received responses from 228 facilities describing 629 surface impoundments [34]. Two hundred were rated using the National Inventory of Dams criteria: 25% were given a High Hazard Potential rating, the worst; 36% were given a Significant Hazard Potential rating; 36% were given a Low Hazard Potential Rating; and 4% were given a Less than Low Hazard Potential Rating. Many of the dams were surprisingly high: 80 units were higher than 50 feet and 133 units were between 25 and 51 feet high. From these data it would seem reasonable to conclude that there is still a high risk for additional large spills with corresponding risks for environmental and health impacts.

The Resource Conservation and Recovery Act (RCRA) is a 1976 law passed by the US Congress that provides general guidelines concerning the disposal of waste (42 USC §6901 et seq.) This law requires the EPA to develop regulations concerning the classification and disposal of various wastes, including coal ash. At the present time it is classified as an exempt waste and there are no constraints concerning its disposal [35]. Because of mounting concerns about heavy metal and other toxic components in coal ash, how it is stored, and how storage may affect health and the environment, the EPA is in the process of making a rule to regulate these practices [36]. The Agency is collecting comments and has held a series of contentious public meetings at which individuals, organizations, and companies delivered testimony. The Agency proposes two alternatives, known as Subtitle C and D. Table 4.4 outlines some of

Table 4.4
Key provisions of proposed rule regulating coal combustion residuals [36]

	Subtitle C	Subtitle D
Enforcement	State and federal	Citizen suits, states are citizens
Corrective action	Monitored by authorized states and EPA	Self-implementing
Permit issuance	Federal requirement for issuance by states	No
Storage requirements	Yes	No

the key differences between these options. Both require groundwater monitoring at sites adjacent to landfills. Provisions outlined in Subtitle C are more stringent than those in Subtitle D. Predictably, environmentalists favor Subtitle C and industry favors Subtitle D. Health considerations, cost, and "unwarranted regulatory interference" are focal points in the debate. Industry is concerned that if coal ash is classified as a hazardous waste, secondary uses, such as inclusion in wallboard and cement, will suffer.

References

1. Kingston Fossil Plant, 2010. Available at http://www.tva.gov/sites/kingston.htm.

2. US Energy Information Administration. Electricity Generating Capacity. Washington DC: Government Printing Office, 2010.

3. US Environmental Protection Agency. Clean Air Markets—Data and Maps, 2008 Emissions, 2008. Available at http://www.epa.gov/airmarkets.

4. Paybins KS, Messinger T, Eychaner JH, Chambers DB, Kozar MD. Water quality in the Kanawha–New River Basin: West Virginia, Virginia, and North Carolina, 1996–98. Circular1204. US Department of the Interior, US Geological Survey, 2010.

5. China Labor Bulletin. Deconstructing deadly details from China's coal mine safety statistics, 2006. Available at http://www.clb.org.hk/en/node/19316.

6. Injuries, Illnesses, and Fatalities in the Coal Mining Industry, April 2010. Available at http://www.bls.gov/iif/oshwc/osh/os/osar0012.htm.

7. Centers for Disease Control. Pneumoconiosis prevalence among working coal miners examined in federal chest radiograph surveillance programs—United States, 1996–2002. MMWR—Morbidity and Mortality Weekly Report 58(50):1412–6, 2003;52(15):336–40.

8. Laney AS, Attfield MD. Coal workers' pneumoconiosis and progressive massive fibrosis are increasingly more prevalent among workers in small underground coal mines in the United States. Occup Environ Med 67(6):428–31, 2010.

9. Seaton A. Coal workers' pneumoconiosis in small underground coal mines in the United States. Occup Environ Med 2010;67(6):364.

10. Huang X, Li W, Attfield MD, Nadas A, Frenkel K, Finkelman RB. Mapping and prediction of coal workers' pneumoconiosis with bioavailable iron content in the bituminous coals. Environ Health Perspect 2005;113(8):964–8.

11. McCunney RJ, Morfeld P, Payne S. What component of coal causes coal workers' pneumoconiosis? JOEM 2009;51(4):462–71.

12. West Virginia Division of Culture and History. Buffalo Creek. West Virginia Archives and History, 2010.

13. Ecology and Environment I. Martin County Coal Corp. Coal Slurry Release Work Plan. Lancaster NY, 2001.

14. Lovan D. Inez coal slurry spill: toxic sludge from Massey facility still pollutes Kentucky town a decade after disaster. Huffington Post, 2010 Oct 10.

15. Coal Impoundment Project WJU. Coal Impoundment Location and Information System, 2010. Wheeling Jesuit University.

16. Committee on Coal Waste Impoundments, Committee on Earth Resources, Board on Earth Sciences and Resources. Coal Waste Impoundments: Risks, Responses and Alternatives. National Research Council, National Academy of Sciences, 2002.

17. Smith V. Critics question safety of storing coal slurry. Associated Press, 2009 Mar 21.

18. Bunnell JD, Bushon RN, Stoeckel DM, et al. Preliminary geochemical, microbiological, and epidemiological investigations into possible linkages between lignite aquifers, pathogenic microbes, and kidney disease in northwestern Louisiana. Environmental Geochemistry and Health 2006;28(6):577–87.

19. Stefanovic V, Cukuranovic R, Miljkovic S, Marinkovic D, Toncheva D. Fifty years of Balkan endemic nephropathy: challenges of study using epidemiological method. Renal Failure 2009;31(5):409–18.

20. US Department of Transportation, US Department of Commerce. 2007 Commodity Flow Survey. Washington DC: Government Printing Office, 2010.

21. US Environmental Protection Agency. Control of Emissions of Air Pollution From Nonroad Diesel Engines and Fuel; Final Rule. Washington DC: EPA, 2010.

22. US Department of Transportation. Accident/Incident Overview, Railroad, January to September, 2010. Washington DC: Government Printing Office, 2010.

23. National Highway Traffic Safety Administration. Highlights of 2009 Motor Vehicle Crashes. Washington DC: Government Printing Office, 2010.

24. Goodell J. Big Coal: The Dirty Secret behind America's Energy Future. Boston: Houghton Mifflin, 2006.

25. US Department of Energy. Fluidized Bed Technology—Overview, 2010. Available at http://www.fossil.energy.gov/programs/powersystems/combustion/fluidizedbed_overview html.

26. Committee on Mine Placement of Coal Combustion Wastes. Managing Coal Combustion Residues in Mines. Washington DC: National Academies Press, 2006.

27. US Energy Information Administration. International Energy Outlook 2010. Washington DC: Government Printing Office, 2010.

28. Stanislaus M. Testimony of Mathy Stanislaus. 7–28–2009. Committee on Transportation and Infrastructure, US House of Representatives, Subcommittee on Water Resources and the Environment.

29. Lisenby D, Tuberty S, Babyak C. Results of ICP-OES Analyses of the TVA Ash Spill Samples Collected on 12–27–08 from the Emory River. 2010.

Available at: http://www.appvoices.org/resources/Preliminary_TVA_Ash_Spill _Sample_Data_AppVoices_December%202008.pdf.

30. US Geological Survey. Radioactive Elements in Coal and Fly Ash: Abundance, Forms, and Environmental Significance Fact Sheet FS-163–97. Washington DC: USGS: 1997.

31. Prepared for EPA by RTI International. Human and ecological risk assessment of coal combustion wastes: Draft. Research Triangle Park NC, 2007.

32. US Environmental Protection Agency Office of Solid Waste. Coal Combustion Waste Damage Case Assessments. Washington, DC: EPA, 2007.

33. US Environmental Protection Agency. Coal Combustion Residuals, 2009. Available at http://www.epa.gov/waste/osw/nonhaz/industrial/special/fossil/ coalashletter htm.

34. US Environmental Protection Agency. Responses from Electric Utilities to EPA Information Request Letter, 2010. Available at http://www.epa.gov/osw/ nonhaz/industrial/special/fossil/surveys/index.htm.

35. US Environmental Protection Agency. Hazardous Waste Regulations, 2011. Available at http://www.epa.gov/osw/laws-regs/regs-haz.htm.

36. US Environmental Protection Agency. Coal Combustion Residuals—Proposed Rule, 2010. Available at http://www.epa.gov/osw/nonhaz/industrial/special/ fossil/ccr-rule/index.htm.

5
Mitigation of Pollutants from Burning Coal

"Fumifugium: The Inconvenience of the Aer And Smoak of London. Together with some Remedies Humbly Proposed by J.E. Esq; To His Sacred Majestie and to the Parliament now Assembled."
—The title of John Evelyn's treatise on coal and its pollutants, 1661 [1]

When Edison's Pearl Street Power Station went on line in 1882 all of the pollution from the coal boilers went straight into the air. Times have changed. Thanks largely to the Clean Air Act, passed by Congress in 1963 and amended significantly in 1970 and 1990, the days of wanton discharges of pollutants into the air are over. Air quality is improving steadily due to regulations that have mandated the installation of pollution control devices at power plants, requirements for cleaner fuels, and other measures. This chapter will describe some of the technologies currently employed to perform this task.

The chapter is divided into sections that deal with carbon dioxide and other pollutants. The rationale for this division is based on the vast differences in the technologies that are required.

Pollutants Other Than Carbon Dioxide

The task of reducing the emission of pollutants created by burning coal arguably begins at most mines where coal is washed or, in the jargon of the World Coal Association, it undergoes "coal beneficiation or coal preparation," before it is shipped [2]. As described in chapter 4, this step removes some impurities, such as sulfur, that contribute to coal-related pollution. Other technologies, such as fluidized bed combustion, optimize the conditions under which coal is burned, improving efficiency and controlling combustion temperatures, an import factor in the production of NO_x and SO_3 emissions. This section deals with the technologies that are employed between the combustion chamber and the tip of the stack.

The challenges associated with the removal of pollutants center on their ability to treat as much as 2.5 million cubic feet of flue gas per minute and the costs of the technology.

Control of Particulates

The particulate matter suspended in the hot gases that leave the combustion chamber is known as fly ash. Most of this ash is composed of silicon dioxide and calcium oxide. These compounds are abundant in coal and are not removed completely during pre-combustion washing. In addition fly ash contains arsenic, beryllium, boron, cadmium, chromium, chromium VI, cobalt, lead, manganese, mercury, molybdenum, selenium, strontium, thallium, thorium, uranium, and vanadium, all of which are present in the coal itself. In addition there are compounds formed during combustion, including dioxins (and other dioxin-like compounds that are lumped in with dioxins) and polycyclic aromatic hydrocarbons.

Although some of these elements are contained within ash particles, some are present as gases, along with oxides of sulfur and nitrogen. As described more completely in chapter 3, these are primary particulates—those formed directly during combustion. Post-combustion particle removal technologies are designed to remove these particles.

Electrostatic precipitators are one of the most commonly employed technologies used to remove particulates. Detailed descriptions of these devices can be found on the web pages maintained by manufacturers, such as Hamon Research–Cottrell. Commercial-scale electrostatic precipitators were patented by Frederick Cottrell over one hundred years ago. The technology is relatively simple. The flue gases pass between electrodes that impart an electrical charge on the particles. A large voltage difference is required for efficient operation. The charged particles are attracted to a collecting electrode where their charge makes them "stick." As the charge dissipates, "rappers" tap on the collecting electrode assemblies causing the particles to fall into hoppers where they can be collected and disposed of.

Electrostatic precipitators become less efficient as the particle size increases. For these larger particles fabric filters, or baghouses, offer a more cost-efficient solution. These filters are similar to the filters typically found on vacuum cleaners used in the home. The particle-laden flue gas enters the baghouse and is forced through a fabric filter. The filters are usually made up of synthetic fibers supported by "scrims" composed of other fibers that have the high mechanical strength required to support the filter. The filter is exposed to harsh conditions including mechanical, thermal, and chemical stress [3]. Clearly, proprietary differences exist among the available options. During operation the particles trapped on the fibers form a "dust cake." The dust cake decreases the effective size of the pores in the baghouse and enhances the efficiency of the filtration process. This also increases the amount of energy required to force gases through the filter. Therefore the filters must be cleaned periodically. As with electrostatic precipitators, the dust cake is collected in hoppers for disposal. Baghouse filters are replacing electrostatic precipitators at many locations because of increasing emission control requirements and their compatibility with desulphurization systems that remove sulfur dioxide [3].

Control of Sulfur Oxides

Removal of sulfur dioxide from coal plant emissions began in earnest in England when the House of Lords upheld the claim of a landowner that SO_2 was damaging his land [4]. The first significant mitigation system was installed in 1931 at the Battersea Station [4]. This requirement spread to include other power plants in London [4]. The onset of World War II put an end to this mandate [4]. The 1970 amendments to the Clean Air Act led the EPA to promulgate regulations that, along with additional amendments related to the Acid Rain Program, have led to progressive restrictions on the emission of sulfur oxides [5].

The removal of sulfur oxides occurs primarily in the second stage of flue gas purification, after the removal of fly ash particulates by electrostatic precipitators or baghouse filters. The removal depends on chemical reactions between sulfur oxides and calcium carbonate from limestone or lime. The reactions are, respectively,

$$CaCO_3 + SO_2 \rightarrow CaSO_3 + CO_2$$

or

$$Ca(OH)_2 + SO_2 \rightarrow CaSO_3 + H_2O$$

In a measure designed to reduce costs, some facilities subject the calcium sulfite ($CaSO_3$) to further oxidation to form gypsum ($CaSO_4$):

$$CaSO_3 + H_2O + \tfrac{1}{2}O_2 \rightarrow CaSO_4 + H_2O$$

These scrubber systems may be one of several design types that vary substantially in their efficiency [6]. According to the EPA, approximately 85% are so-called wet systems, which are the most efficient, but the most expensive. Wet systems are not well suited to the retrofitting needed to improve sulfur oxide reductions in older plants. Wet systems typically remove 90% or more of the SO_2 from flue gases. In these systems a slurry of limestone or lime is injected into the flue gas stream. There are several mechanisms for removal of the precipitate that depend on the weight of the precipitate compared to other flue gases. Dry systems are substantially less efficient. Typically they remove 80% or less of the SO_2. They account for approximately 3% of installed systems. An intermediate semi-dry or spray dry system makes use of aqueous slurry that contains less water than the wet systems. These account for the remaining 12% or so of systems. As with the dry systems, the sorbent, or SO_2 trapping chemical, is injected before the flue gas reaches the electrostatic precipitator or baghouse filter that traps the particles formed.

The gypsum that is formed by forced oxidation reactions associated with the wet method also contains mercury and other metals that are in coal combustion wastes. According to a report prepared for the US Department of Energy, over 7.5 million tons of gypsum from coal plants was used to make wall board in 2006 [7]. This wallboard may contain 0.01 to 0.17 grams of mercury per ton of gypsum. This mercury is in the elemental form (Hg^0) rather than the methylmercury that is formed by bacterial action on elemental mercury in aqueous environments. Nevertheless, this contaminant is a source of concern to many scientists and is a factor in the debate concerning the classification of coal combustion waste, or coal ash.

As pollution-control regulations become more stringent and more pollutants are removed at coal plants, the volume of the waste increases and the amount of toxic metals and other compounds in the waste increases. This paradoxical increase in toxicity poses conceptual, ethical, and regulatory issues. Should coal ash be classified as a hazardous waste or not? Should coal plant owners be permitted to dispose of these wastes as they see fit, or should regulations be imposed that are designed to prevent exposure of the public to this increasingly toxic by-product?

Carbon Dioxide Capture and Storage

According to the Intergovernmental Panel on Climate Change Special Report on Carbon Dioxide Capture and Storage, this is "a process consisting of the separation of CO_2 from industrial and energy-related sources, transport to a storage location, and long-term isolation from the atmosphere" [8]. Carbon capture and storage is potentially a technological fix to prevent the climate change threat posed by the release of CO_2 associated with burning fossil fuels. In essence, if carbon capture and storage worked perfectly, we could burn coal and not cause global warming. Indeed, there are organizations such as the American Coalition for Clean Coal Electricity that advocate the "robust utilization of coal." As an indicator of the underlying controversies surrounding clean coal and climate change, Duke Energy, a major supplier of coal-produced electricity with assets of $53 billion, resigned from this coalition in September of 2009. Their decision was based on the failure of others in the industry to support congressional measures designed to address climate change.

In a very real sense, the use of the term "capture" does not characterize the process properly. In the carbon capture and storage usage, capture means separation or production of highly purified CO_2 that is the consequence of burning a carbon-based fuel, such as coal or natural gas in an oxygen-containing atmosphere. In more conventional usage, capture implies the act of taking a prisoner, namely placing a criminal in captivity where he or she no longer poses a threat to society. It is possible to "capture" CO_2 in a coal-fired power plant and promptly release the gas into the atmosphere. This, of course, does not help mitigate the effects of producing CO_2. Sequestering the gas is essential to complete the process.

According to the Intergovernmental Panel on Climate Change's Special Report, coal-fired electrical power plants generate approximately 59.59% of all CO_2 released worldwide by stationary sources [8]. They identified 4,942 sources that released 10,539 million metric tonnes of CO_2. Although more coal is burned in Texas than any other state, as shown in figure 1.2, many of these sources in the United States are located in the midwestern and eastern part of the country (as shown in figure 13.2). Other major sources are located wherever coal is burned: including the northwestern part of Europe, eastern China, and the Indian subcontinent.

꘎   It is important to remember that none of the carbon capture and storage methods succeed in the elimination of more than 80% to 90% of the CO_2 produced. Thus, even if all new coal-fired power plants were equipped with carbon capture and storage technology, the amount of CO_2 released into the atmosphere by new and existing facilities would continue to increase. The rate of increase would be slowed, however.

Although there are exceptions [8–10], health considerations are not an important aspect of reports describing the carbon capture and storage process. The major focus of most publications rests on the physical and chemical processes, and their costs, that are necessary for large-scale deployment of carbon capture and storage. Although health is the focus of this report, it is necessary to understand some of the basic elements of carbon capture and storage before arguments can be made concerning health.

Carbon Capture

Most of the sources emitting large amounts of CO_2 produce a waste stream that contains less than 15% of the gas—fewer than 2% produce emissions that are at least 95% CO_2 [8]. Thus carbon capture is a formidable task. There are three basic strategies that are used to capture the CO_2 produced in electric utilities: post-combustion and pre-combustion capture, and the oxyfuel combustion.

Post-combustion processes typically employ a liquid, such as monoethanolamine, or some other aqueous amine (shown as RNH_2 in the equation below where R is typically an organic moiety), to form a soluble carbonate salt.

$$2RNH_2 + CO_2 + H_2O \rightleftarrows (RNH_3)_2CO_3$$

This reaction is reversible, and after cooling, the absorbed CO_2 is stripped away from the amine in a second stage, thereby regenerating the amine and yielding a stream of purified CO_2. Energy costs in this process are largely related to the production and replacement of the amine, cooling the flue gas, regeneration of the amine, and compression and liquification of the CO_2 prior to transport and storage.

Precombustion capture systems also employ a two-stage process. In the first stage, the coal (or other fuel) is reacted with steam and air or oxygen to produce a mixture of carbon monoxide and hydrogen gas. In the second stage, in the so-called shift reactor, steam is injected into the hydrogen carbon monoxide mixture to form additional hydrogen gas and CO_2. Separation of this mixture into separate gas streams of hydrogen and CO_2 completes this initial phase. The resulting hydrogen gas can then be burned to produce electricity. It can also be used in fuel cells, such as those proposed for use in the transportation industry to replace fossil fuels. The plants for which this technology is most applicable are those incorporating integrated gasification combined cycle technology. This technology is applicable to both coal and natural gas fueled plants. Costs are related to capital expenditures required to build and operate the plant, including the requisite increases in energy use.

The production of oxygen gas with a 95% or higher purity is the initial step in oxyfuel-based capture systems. By burning coal in a high-oxygen environment, the flue gas consists mainly of CO_2, water, nitrogen, and other pollutants, such as sulfur dioxide, nitrogen dioxide, particulates, and mercury. Cooling and compression are required to remove water vapor. Additional separation techniques may be required to remove nitrogen and other compounds before the CO_2 is transported to a storage site. Costs of oxyfuel systems are those associated with large-scale separation of oxygen from air, capital costs, and the requisite energy expenditures.

The Intergovernmental Panel on Climate Change report estimates that the typical costs for CO_2 capture, excluding transportation and storage, are on the order of 4 to 9 cents per kWh [8]. This estimate applies to newly constructed power plants designed and built to perform carbon capture and storage. Reliable cost estimates for retrofitting existing power plants to enable carbon capture and storage are not available from that source.

Carbon Dioxide Transport

Unless a point source for CO_2 generation is located on the same site as that chosen for storage, it is necessary to transport the gas. This is done most efficiently after the gas is compressed and converted to a liquid, which makes CO_2 easier and more economical to transport and store at various stages before storage. There are several mature technologies currently in use that are suitable for CO_2 transport including pipelines, rail, and ship.

In the United States about 3,600 miles of pipeline are in place to transport CO_2 [11]. Approximately 50 million tons of CO_2 are transported via these pipelines each year. These pipelines are located mainly in Texas, where the CO_2 is used to enhance the recovery of oil from wells. For various technical reasons, related to corrosion of pipelines, it may be necessary to purify the CO_2 so that it is more suitable for transport. This is probably not an insurmountable problem. Liquids that have properties similar to liquified CO_2 are already transported via pipelines through many varied ecosystems, including deserts, mountains, and oceans as deep as 2,200 meters. At present, the Intergovernmental Panel on Climate Change report indicates that approximately 10,500 million tons of CO_2 are produced per year by almost 5,000 sources. Thus, to accommodate all of this CO_2, pipeline capacity would need to increase by a factor of over 200, ignoring the fact that additional coal-fired power plants will be built. If carbon capture and storage technology becomes a reality, it is likely that power plants close to suitable storage sites will be among the first to use it.

Since liquid CO_2 has physical properties that are similar to liquefied natural gas and propane, and since substantial amounts of these two products are transported currently by ship and by rail, it is possible that some CO_2 transport may occur by these routes. However, neither of these options is as inexpensive as transport via pipeline. In addition to adding to rail stock and building many huge ships, additional facilities for loading and off-loading CO_2 would be required.

The Climate Change report includes cost estimates for transporting CO_2. For distances up to and including approximately 1,500 kilometers, onshore and offshore pipelines and marine transport costs are projected to be very similar. As distances increase, pipeline costs rise linearly with a steeper slope for the more expensive offshore systems, while marine transport rises more slowly and nonlinearly. For distances of about 3,000 kilometers, transportation by ship is about half the cost of offshore pipeline transport and below that of onshore transportation. Assuming a distance of 1,500 kilometers, where the three possibilities are likely to cost about the same amount, an annual CO_2 production of 10,500 million tons of CO_2 each year, and transportation costs of US\$17 per ton, transportation costs alone would be approximately US\$18 billion each year. This cost would be phased in as carbon capture and storage sites come on line.

The Secretary of Transportation has the primary authority to regulate interstate CO_2 pipelines under the authority of the Hazardous Liquid

Pipeline Act of 1979 [11]. However, there are numerous regulatory hurdles that must also be surmounted before the large-scale transportation of CO_2 becomes a reality. CO_2 that is used to enhance oil recovery is a commodity, and as such, it is subject to regulation by states and the Bureau of Land Management. Since enhanced oil recovery processes will use only a limited amount of the vast quantities of CO_2 that are generated by burning coal, most of it will need to be disposed of as an industrial pollutant, where it comes under the authority of the EPA. Recent US Supreme Court decisions have strengthened the ability of the EPA to regulate CO_2 under the Clean Air Act. To add another complication, the EPA also plans to regulate deep-well injections of CO_2 under the authority given to it by the Safe Drinking Water Act because of the threat that these injections may affect ground water. It is virtually certain that Congress will need to act to provide clarity.

Isolation of Carbon Dioxide from the Atmosphere

The final critical step in carbon capture and storage in the terms of the Intergovernmental Panel on Climate Change definition, is "long-term isolation from the atmosphere." The options for storage are many and varied [8]. They are all controversial. The Panel's Special Report on carbon capture and storage includes separate chapters on underground storage, deep ocean storage, and mineral carbonation process. These chapters make up almost half of the report.

Underground storage of CO_2 has received a great deal of attention. This is also the technique for which there is the most experience. Relatively widespread use of CO_2 in oil recovery projects and at least three commercial projects using this technology form the basis for advocating this strategy.

The first step is the identification of a suitable disposal site. Typically these sites are deep wells drilled through a caprock that is impermeable to CO_2 into a deep rock formation. At depths below 1,000 meters, pressures are such that CO_2 has many properties of a liquid. The injected CO_2 is then trapped in pores of the rock, dissolved in liquid phase elements in the formation, such as water, or absorbed onto organic matter, such as coal that is present in the rock (often shale). By avoiding known fault lines, existing wells, and so forth, the authors of the Intergovernmental Panel Special Report predict that "carefully selected sites can store [CO_2] underground: it is considered likely that 99% or more of the injected CO_2 will be retained for 1,000 years" [8]. Any CO_2 that escapes through the caprock is likely to react with minerals in the formation to form insoluble carbonates, enhancing the probability that the CO_2 will remain isolated from the atmosphere. CO_2-enhanced oil recovery may reduce the cost of underground storage because significant infrastructure is already present at these sites and the additional oil that is recovered will offset some of the costs. The storage capacity of reservoirs beneath oil and gas wells is estimated to be 650 to 900 billion tons of CO_2. However, the relative paucity of suitable sites may be a limiting factor. Widespread deep saline reservoirs are likely to be highly favored disposal sites. One trillion tons or more of CO_2 storage capacity in brine formations appears to be likely.

The cost estimates for underground storage range between 0.6 and just over \$US8 per ton of CO_2 [8]. The wide range of these estimates arises from differences in the characteristics of various sites. Concerns that arise from this proposed method relate to uncertainties about the viability of storage sites that could lead to significant releases of CO_2 into the atmosphere or contamination of water or both. Careful on-site monitoring will be required and regulations governing the process will need to be promulgated and enforced.

Deep ocean storage of CO_2 will undoubtedly prove to be the most controversial of the proposed isolation strategies. Much of the rationale for oceanic storage is based on theory and modeling with little in the way of practical demonstration in pilot studies.

The intentional injection of CO_2 into oceans should be seen as distinct from the passive oceanic storage of atmospheric CO_2 that is currently taking place [12]. It is thought that about 500 billion tons of CO_2 have dissolved in the upper portions of the world's oceans since the advent of the industrial revolution (out of approximately 1,300 billion tons of CO_2 that have been produced). This has resulted in an acidification of surface waters with a drop in pH of about 0.1 unit. Oceanic uptake of CO_2 and further acidification are almost certain to continue or accelerate as atmospheric CO_2 concentrations rise. This seemingly small drop in pH, along with ocean warming, is thought to underlie some of the observed changes in marine ecosystems, such as loss of biodiversity, reduced rates of calcification of corals, and impaired reproduction and growth of some marine species.

The physical properties of CO_2 make this method for its storage worth theoretical consideration. Below depths of 3,000 meters, pressures are high enough to maintain CO_2 in a liquid phase with a density greater than water. Liquid CO_2 injected below this level from pipelines or fixed platforms will form undersea "lakes" in undersea valleys. This CO_2 is expected to remain in situ for very long periods of time. Since oceans cover some 70% of the earth's surface and have an average depth of approximately 3,800 meters, the upper bound for the potential capacity for intentional injection of CO_2 into the oceans is very high.

Since this disposal strategy exists mainly in the theoretical realm, there are insufficient hard data to make accurate risk assessments and to predict the ecological consequences of intentional deep ocean CO_2 disposal. Some consequences are certain. Naturally occurring marine life would be exterminated in underwater CO_2 lakes. It is likely that there would be acidification of the water at the water–CO_2 boundary that would also have adverse effects on marine life. Eventually, in a geological time frame, there would be mixing of CO_2-enriched water at boundary layers with upper oceanic water. Without more information about affected ecosystems, it is impossible to predict the effects of these disruptions and loss of biodiversity. On the other hand, the predicted time scale for equilibration of deep ocean CO_2 with surface water is very long, and thus this aspect of proposed CO_2 disposal probably does not pose an immediate threat. However, the major reason for considering carbon capture and storage technology is to get past short- and intermediate-term effects of atmospheric CO_2 releases and to consider and manage the long-term consequences.

Before large-scale, or perhaps even small-scale, pilot injections of CO_2 into the ocean could be undertaken, binding international agreements would be necessary, public opposition would need to be dealt with, and well-designed monitoring programs would need to be in place to properly assess the impact of CO_2 on marine ecosystems.

Mineral carbonization, or mineral storage, as it is sometimes called, is the third process that has been proposed to isolate CO_2 from the atmosphere. In the process of mineral carbonization, CO_2 is reacted with metal oxides, such as oxides of calcium and magnesium, to form carbonates. These carbonates are stable, and environmentally acceptable disposal sites should not require monitoring to detect any escape of CO_2. Some rocks, such as serpentine and olivine, are potential candidate sources for these oxides along with slag from the production of steel and fly ash from coal-fired power plants. Olivine and serpentine are both abundant rocks. The carbonization reaction could be carried out by injecting CO_2 directly into the rock formation or at the site of the power plant after extraction of the mineral. The high costs of carbonation, associated with mineral extraction and the use of energy to force the chemical reaction to completion in a reasonable time frame, make this a relatively unattractive option. The Intergovernmental Panel on Climate Change Special Report's cost estimates for a carbonization-equipped plant range between 60% and 180% of the cost of a conventional plant without carbon capture and storage technology.

The Calera Company has built a demonstration project at Moss Landing, CA, designed to sequester 30,000 tons of CO_2 per year [12]. The major chemical reactions for the process are basically those of mineral carbonation:

$$Ca^{++} + 2OH^- + CO_2 \rightarrow CaCO_3 + H_2O$$

$$Mg^{++} + 2OH^- + CO_2 \rightarrow MgCO_3 + H_2O$$

The alkali sources include fly ash, waste water, and brines. In a 2010 *New York Times* column, Thomas L. Friedman touted the process and included seawater in the list of brines, noting the large volume of the oceans and the ability of corals to form carbonates [13]. In addition to CO_2, the process consumes hydroxyl ions, meaning it acidifies the waste stream. Large-scale use of ocean water as a source of hydroxyl ions would lead to acidification of water returned to the ocean and the potential for environmental harm.

Adverse Health Effects and Risks Associated with Carbon Capture and Storage

The most obvious threats to health posed by the carbon capture and storage strategies reviewed above would occur in the event of the release of large amounts of CO_2. The potential releases could occur at any of the stages: at the site of CO_2 capture, during transport, or transfer, whether by pipeline, rail, or ship, and so forth; or escape during or after storage (depending on the storage method).

Carbon dioxide is a colorless, odorless gas that is heavier than air. It may cause symptoms or death by displacing oxygen from inhaled air, leading to hypoxia and asphyxiation, or by causing symptomatic or fatal acidification of the blood and body fluids after inhalation. Increases in the partial pressure of CO_2 in arterial blood occur rapidly as the percentage of CO_2 increases in inspired air. This leads to an increase in the respiratory rate, blood pressure, and the rate at which blood flows to the brain. The following effects are observed with increasing concentrations of CO_2: 2% to 5%, headache associated with changes in brain blood flow, dizziness, sweating, shortness of breath; 6% to 10%, rapid breathing, rapid heart rate, increased dizziness; 11% to 17%, drowsiness leading to coma, twitching of muscles; >17%, epileptic seizures, coma, and death [14]. The Occupational Safety and Health Administration has established the permissible exposure level for carbon dioxide in a general industrial setting at 5,000 ppm (parts per million) [15].

The natural disaster at Lake Nyos in Cameroon that occurred on August 21, 1986, provides a glimpse of what a large-scale release of CO_2 could look like [16]. Lake Nyos is located in a volcanic crater. The lake is about 250 meters higher than several small towns to its north. The release took place between 9 and 10 pm and the gas flowed downhill into Nuos, Subum, and Cha, where about 1,700 people died. Survivors reported the smell of rotten eggs and gunpowder, but samples from the lake and an examination of survivors, animals, and plants suggested that the predominating gas was CO_2. Skin lesions and weakness of arms and legs in some of the survivors were attributed to prolonged coma-induced immobilization and damage to nerves and skin. Most of those who regained consciousness survived. Similar but less catastrophic events have been observed at other volcanic areas. It is thought that the total amount of gas released was on the order of 250,000 metric tonnes (2,200 lb per tonne) or about one billion cubic meters. Thus the sudden release of large amounts of CO_2 clearly has the potential to cause a catastrophe. This potential is increased by the physical properties of CO_2: it is heavier than air and therefore likely to accumulate in low-lying areas such as basements, tunnels, and other areas below ground level. It is also odorless, and very high, potentially lethal levels of the gas may not be detected by those at risk.

According to reports from the Office of Pipeline Safety, there were twelve accidents that resulted in the release of CO_2 from pipelines during the two decades between 1986 and 2006 (cited by [11]). No injuries or deaths occurred. Only a very small portion of all pipelines carry CO_2, and during those same decades, there were 5,610 pipeline accidents (natural gas and other hazardous liquids) that killed 107 and injured 520. The Congressional Research Service estimates that CO_2 pipeline accidents are likely to increase in number. As pipelines age, the accident probability is likely to be similar to that observed for natural gas pipelines. It seems inevitable that an accidental leak from a CO_2 pipeline will occur and cause deaths and injuries.

The pipeline industry has a relatively poor safety record. Since 1990 approximately 5,600 pipeline accidents that released a total of 110 million gallons were reported to the Department of Transportation, according to an article published by *The New York Times* [22]. Equipment failure due to faulty construction techniques and installation accounted for almost half of the spills and pipeline corrosion for about a quarter. The industry is largely self-policed. Although the safety record has improved during the last decade, spills still occur at a rate of about one every three days.

The National Transportation Safety Board investigates and files reports involving hazardous substances transported via pipelines. A brief synopsis of an accident that occurred near Knoxville, Tennessee, in February 1999 is informative [17]. The incident began on the afternoon of February 9, 1999, as an apparently successful delivery of diesel fuel took place via a relatively small pipeline. At the completion of the delivery, all the valves in the pipeline were closed, leaving pressurized fuel in the pipe. Monitoring systems recorded a drop in pipeline pressure which was automatically registered at the company's distant Atlanta, Georgia, control center. This pressure drop did not trigger an alarm and went unnoticed by the operator. Just after midnight, a man noted a kerosene odor while walking outside of his Knoxville home. He did not take any action. At 1:02 am, the first of several 911 calls was made after another resident smelled diesel fuel or gasoline. The Knoxville Fire Department responded to the call and concluded that the odor came from an asphalt plant across the river. A second 911 call was placed an hour later and the fire department again concluded that the smell came from the asphalt plant. A third 911 call was made after oil was spotted on the Tennessee River. Meanwhile the Atlanta staff, still unaware of the problem, called and asked the Knoxville operator to begin another delivery of the product. After the expected increases in pipeline pressure did not occur, a senior controller was consulted and a shutdown of the pipeline was initiated. Operators noted that the fuel delivery was short by an atypically large amount. At about 4:00 am a pipeline employee reported that he found no evidence of a leak. Still unaware of a problem, the operators repressurized the system. Low pressures and a delivery rate that was only about two-thirds of that expected were noted. Finally, at about 4:30 am, the Knoxville Fire Department "found diesel fuel spraying onto a . . . residence."

Although there were no deaths or injuries associated with the accident, property damage of about $7 million occurred due to the loss of about 53,000 gallons of diesel fuel. Multiple delays occurred as several individuals or groups failed to interpret data correctly or to find the leak. As is typical of many accidents, a series of events took place that piled one on top of the other. The problem became steadily worse as the leak continued.

In addition to spontaneous, natural releases of carbon dioxide and pipeline accidents, there is growing evidence that underground storage may not be as permanent as some would hope. A recent report in the Winnipeg Free Press may illustrate this potential problem [18]. According to the article, a Saskatchewan farm couple began to notice algae blooms, clots of foam, and scum on ponds in gravel pits on their land in 2005. Dead animals were found a short distance from the ponds that bubbled on occasion. Because of repeated explosion-like booms, fizzing water, and other factors, the couple abandoned their land and hired Petro-Find Geochem to investigate the phenomena. This company specializes in measuring subterranean gases and their relationship to the petroleum industry. The Petro-Find report alleges that concentrations of CO_2 that were as high as 110,607 ppm were found at the epicenter of the high-CO_2 site. The reported CO_2 concentration was as high as 17,000 ppm within a radius of 100 meters of the epicenter [19]. The National Institute for Occupational Safety and Health limit is 5,000 ppm [15].

The Petro-Find report alleges that the CO_2 came from the 6,000 metric tonnes of liquid CO_2 injected daily into the Weyburn oil field to enhance oil recovery. Their contention is based on measurement of the ratio of two nonradioactive isotopic forms of carbon (carbon-12 and carbon-13) in the samples from the farm that were compared to published values for the ratio in CO_2 being injected by the oil-field operators. The carbon isotope ratio for CO_2 arising naturally is quite different from that which is being injected. This was cited as evidence that the injected CO_2 was leaking and was not permanently isolated from the atmosphere.

This report is at odds with an earlier, much more detailed study conducted by the Petroleum Technology Research Center [20]. Investigators from this center made measurements of the isotopic ratio that were similar to those in the report of the alleged leak. They also used other techniques to look for evidence of leaking carbon dioxide from the Weyburn field. They found no evidence for a leak when they compared pre–CO_2 injection data to data obtained 930 days after the CO_2 injection process began. *The Toronto Globe and Mail* quoted a senior scientist from the PTRC as saying, "For them to claim that it [the isotopic data] is indicative of the injected gas [i.e., the CO_2 injected at the Weyburn site] is not appropriate" [21]. The article concluded by saying that the PTRC official called for a more in-depth study of the phenomena, ". . . if something is there, we certainly want to find it."

References

1. Fumifugium: The Inconvenience of the Aer And Smoak of London. Together with some Remedies Humbly Proposed by J. Evelyn Esq; To His Sacred Majestie and to the Parliament now Assembled. London: Published by His Majesties Command; 1661.

2. The Coal Resource. The World Coal Institute. Available at http://www.world-coal.org. Accessed 2009.

3. Klotz A, Haug B. Experiences in bag house applications after coal fired boilers. Paper 6C2. International Conference on Electrostatic Precipitation X, 2006.

4. Biondo SJ, Marten JC. A history of flue gas desulfurization systems since 1850: Research, development, and demonstration. J Air Pollution Control Assn 1977;27(10):948–61.

5. Rogers PG. The Clean Air Act of 1970. EPA Journal 1990. Available at http://www.epa.gov/aboutepa/history/topics/caa70/11.html.

6. US Environmental Protection Agency. Air Pollution Control Technology Fact Sheet EPA-452/F-03–034. Washington DC: EPA, 2011.

7. Sanderson J, Blythe GM, Richardson M. Fate of mercury in synthetic gypsum used for wallboard production. Pittsburgh: National Energy Technology Laboratory, 2008.

8. Working Group III of the Intergovernmental Panel on Climate Change. IPCC, 2005: IPCC Special Report on Carbon Dioxide Capture and Storage. New York: Cambridge University Press, 2005.

9. Fogarty J, McCally M. Health and safety risks of carbon capture and storage. JAMA 2010;303(1):67–8.

10. Markandaya A, Wilkinson P. Energy and health 2: electricity generation. Lancet 2007; 370:979–90.

11. Parfomak PW, Folger P. Carbon Dioxide (CO_2) Pipelines for Carbon Sequestration: Emerging Policy Issues. Congressional Research Service. Washington DC: Government Printing Office, 2007.

12. The Calera Corporation. Calera: Sequestering CO_2 in the Built Environment. Available at htt;://www.calera.com. Accessed 2011.

13. Friedman TL. Dreaming the possible dream. New York Times, 2010 Mar 10.

14. Langford NJ. Carbon dioxide poisoning. Toxicol Revs 2005;24(4):229–35.

15. Chemical Sampling Information: Carbon Dioxide. Occupational Safety and Health Administration. Available at http://www.osha.gov/dts/chemicalsampling/toc/chmn_C.html.

16. Baxter PJ, Kapila M, Mfonfu D. Lake Nyos disaster, Cameroon, 1986: the medical effects of large scale emission of carbon dioxide? BMJ 1989;298(6685): 1437–41.

17. Pipeline Accident Brief, Knoxville, TN, Accident DCA99-MP005, http://www.ntsb.gov/doclib/reports/2001/PAB0101.pdf. National Transportation Safety Board, 1999.

18. Weber B, Graham J. Land fizzing like soda pop: farmer says CO_2 injected underground is leaking. Winnipeg Free Press, 2011 Jan 11.

19. Lafleur P. Geochemical Soil Gas Survey: A Site Investigation of SW30-5-13-W2M, Weyburn Field, Saskatchewan. Saskatoon, SK: Petro-Find Geochem, 2010.

20. Whittaker S, White D, Law D, Chalaturnyk R. IEA GHG Weyburn CO_2 Monitoring and Storage Project Summary Report 2000–2004. Regina, Saskatchewan: Petroleum Technology Research Centre, 2011.

21. Vanderklippe N. Alleged leaks from carbon storage project questioned. Toronto Globe and Mail, 2011 Jan 11.

22. Frosch D, Roberts J. Pipeline spills put safeguards under scrutiny. New York Times, 2011 Sep 10.

6

Pathophysiology: How Pollution Damages Cells and Tissues

Koyaanisqatsi: from the Hopi language, life out of balance.

This Hopi word provides us with a good description of what happens when people are exposed to coal-derived air pollutants—the normal states of equilibrium that characterize essential biochemical and physiological processes are pushed out of balance by the stresses imposed by pollutants. Imbalance disrupts cellular functions and leads to the production of symptoms, tissue injury, and disease. This is the essence of how pollution leads to disease. The devil is in the details. To avoid confronting the devil, skip to the next chapter!

The air pollution that arises as a result of burning coal is a complex mixture of gases, liquid droplets, and particles, as described in chapter 3. As will be seen in the ensuing chapters, this mix leads to many diverse disorders that range from acute attacks of asthma associated with high ambient ozone levels to the possibility that particulates contribute to the development of neurodegenerative diseases, such as Alzheimer's disease. While the exact mechanisms that cause an exposure to result in a disease are, in many cases, extremely complex and incompletely understood, recent evidence suggests that oxidative stress and inflammation play central roles for many of these.

The Central Role of the Lung

Although there are clear exceptions, inhalation is the primary means by which we are exposed to most air pollutants. This should come as no surprise. The lung is the organ where oxygen is taken up by the body and carbon dioxide is excreted. This carbon dioxide is produced by the metabolism of sugars and other molecules in various body organs. Humans require lots of oxygen and produce lots of carbon dioxide; therefore the surface area of the alveoli, the site of the gas exchange, must be correspondingly large. The volume of air that moves in and out of the lungs must also be large. The interface between alveolar air and the body is where the action begins.

The exposure to airborne pollutants is higher in children than adults. There are several explanations for this increase. According to a review by Bateson and Schwartz, the susceptibility of children to the effects of air pollution is multifactorial and includes the following [1]: (1) Children have different patterns of breathing than adults. (2) They are predominantly mouth-breathers, thereby bypassing the filtering effects of the nasal passages. This allows pollutants to travel deeper into the lungs. (3) They have a larger lung surface area per unit weight than adults. (4) They spend more time out of doors, particularly in the afternoons and during the summer months when ozone and other pollutant levels are likely to be the highest. (5) Children also have higher ventilation rates than adults. They breathe more frequently and breathe more air per unit weight than adults. (6) When active, children may ignore early symptoms of exposure and fail to seek treatment or reduce their exposure by moving to a less polluted environment, such as moving indoors. In addition the diameter of the airways in children is smaller than in adults, and therefore airways may be more susceptible to the effects of the airway narrowing that is characteristic of asthmatic attacks. These factors, combined with the possible adverse impact of pollutants on lung development and the immaturity of enzyme and immune systems that detoxify pollutants, may all contribute to an increase in the sensitivity of children to pollutants produced by burning coal [2].

The alveolar surface is covered by a thin layer of liquid that contains a large variety of compounds. Among these are the sulfur-containing amino acid glutathione (in its reduced state), vitamin C, uric acid, and vitamin E. These chemicals are all antioxidants. They act to maintain the proper state of oxidation that is essential to all cells, by neutralizing oxidizing molecules, such as ozone and N_2O, in the aqueous layer between alveolar air and the cells that make up the alveoli. When the concentration of oxidizing molecules is too high, the capacity of these defenses is exceeded. This leads to an elevation in the concentration of free radicals and the production of oxidative damage, that is, oxidative stress.

Uncontrolled oxidative stress has the potential to initiate a positive feedback cycle that makes a bad situation worse. The increase in the free radicals acts as a trigger that initiates an inflammatory response. The resulting inflammation generates still more free radicals. This leads to an increase in local responses that may be the immediate cause of symptoms, such as an acute exacerbation of asthma. Local responses, in turn, may lead to systemic responses that have the potential to trigger local responses in other organs or other systemic responses, such as an increase in blood pressure. These systemic responses may cause an acute illness, precipitate a hospital admission, or, over a long period, contribute to conditions that take a long time to develop, such as cancer, particularly of the lung.

Reactive Oxygen Species and Oxidative Stress

Reactive oxygen species play important roles in many normal cellular processes. Reactive oxygen species are highly reactive molecules that contain oxygen; that is, they combine readily with other molecules. Hydrogen peroxide and other molecules known as free radicals are forms of reactive oxygen species that occur normally. Free radicals are molecules that contain one or more unpaired electrons. The unpaired electrons are what makes them so reactive. The chemistry of free radicals is fairly complicated. The unique electron configuration of molecular oxygen (O_2, or dioxygen) makes it a radical—the addition of one more electron results in the formation of the superoxide anion radical, a form of oxygen that is even more reactive than dioxygen. The superoxide anion radical is, in turn, converted to hydrogen peroxide in the presence of the enzyme superoxide dismutase. Hydrogen peroxide is a highly reactive molecule, capable of attacking and damaging cellular elements. In the presence of iron, either as Fe^{2+} and Fe^{3+}, various chemical reactions lead to the formation of the hydroxyl radical (the neutral form of the hydroxide ion, OH^-), hydrogen peroxide, and the superoxide anion radical. These highly reactive radicals attack various molecules such as lipids on cell membranes, lipids, proteins, and DNA. The attack on DNA is particularly troubling, as this is a likely initial step in oncogenesis—the production of cancers by environmental pollutants. Excessive concentrations of any of these radicals has the potential to damage cells.

Signal transduction is one of the processes that depends on ROS (reactive oxygen species). Signal transduction is the term used to describe the processes by which "information" or signals, in the form of hormones, growth factors, or other small proteins, such as cytokines, enter cells and act to control a cellular function. In the brain, cytokines are small molecules produced by glial cells that help mediate inflammation. Normal cellular growth and differentiation are two of the processes that depend on ROS. When signal transduction goes awry, normal functions of cells are disrupted. When the concentration of ROS is too high, cells suffer from a condition referred to as oxidative stress. This continuum of increasing ROS concentrations and their role in cause and effect responses moves from normal functions to abnormal or pathological functions through several "tiers," as shown in figure 6.1. These processes are described in great detail in the references provided [3,4].

As the level of oxidative stress increases, cells move from the realm of normal function to a condition where so-called phase II enzymatic processes are triggered. Phase II enzymes are those involved in combining two compounds, for example, combining a pollutant with some compound produced by the cell that is a part of the body's self-maintenance program. This is a process that is common in the detoxification of a variety of compounds, such as drugs and products of cellular metabolism. Inflammation occurs when phase II processes are overwhelmed by additional stress, such as that caused by increases in the concentration of ROS. Severe inflammation kills cells by one of two processes known as apoptosis (programmed cell death) or necrosis.

Apoptosis is a normal phenomenon that is triggered at critical times during the course of development of organs, such as the brain. Apoptosis should be differentiated from necrosis, the type of cellular death that occurs as the result of trauma or an infarct of the heart or brain. Necrosis "clogs" the organ with unwanted debris, whereas apoptosis does not. At birth and soon thereafter, there are many more connections among the neurons than are present when the brain reaches its ultimate stage of development. During the process of development, controlled apoptotic reactions "trim" unwanted and unnecessary neuronal connections and peak efficiency is reached. Thus controlled apoptosis is normal, desirable, even essential to normal development. However, when an abnormal cellular environment triggers apoptosis at an inappropriate

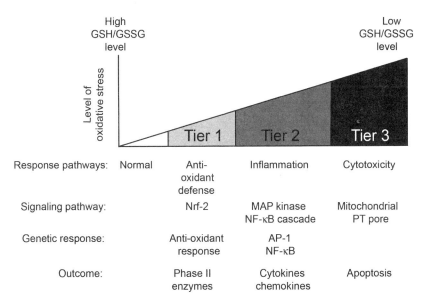

Figure 6.1
Hierarchical model of oxidative stress. At low (tier 1) levels of oxidative stress phase II enzymes are induced in an attempt to restore the oxidative state to more normal levels. At tier 2 levels of stress pro-inflammatory responses are induced. Tier 3 levels are sufficient to affect pores in mitochrondia and cause apoptosis or necrosis. Abbreviations: Nrf-2, nuclear factor (erithroid-derived) like 2, responds to various extracellular stimuli, a master controller of antioxidant responses; MAP kinase, mitogen-activated protein kinase, responds to extracellular stimuli; NF-κB, nuclear factor kappa–light-chain enhancer of activated B cells, controls DNA transcription; AP-1, activator protein 1, a regulator of gene expression. These molecules mediate various elements of the response to oxidative stress. From Nel A, et al., Science 2006;311(5761):622–7. Reprinted with permission from AAAS.

time, normal development is affected and abnormalities of function are likely to occur.

Although oxidative stress caused by excessive levels of reactive oxygen species is now believed to be a contributing factor, or even the primary mechanism that leads to a large number of diseases, it is still a relatively new concept. The term "oxidative stress" appears to have been coined about twenty years ago [5].

As shown in figure 6.2, oxidative stress and inflammation in the lung have systemic effects. These effects have many forms. In their update on the effects of particulate matter on the cardiovascular

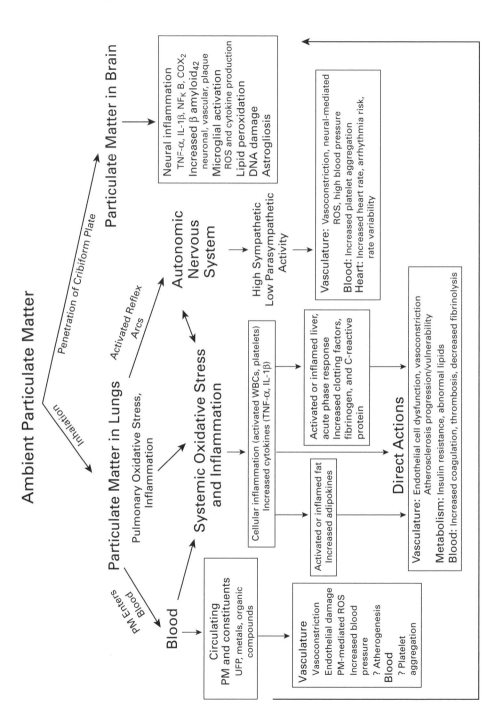

Figure 6.2
Pathways linking cardiovascular and nervous system disease to particulate exposure. Particulates cause acute, subacute, or chronic effects after they enter the blood, produce systemic oxidative stress and inflammation, and affect the autonomic nervous system, and enter the brain via the cribiform plate or the cerebral circulation. Abbreviations: UFP, ultrafine particles; TNF-α, tumor necrosis factor alpha; IL-1β, interleukin 1-beta; WBC, white blood cell, NF$_\kappa$ B, nuclear factor kappa-light chain enhancer of activated B cells; COX$_2$, second isoenzyme of cyclooxygenase; β amyloid$_{42}$, a 42 amino acid protein associated with Alzhemier's disease. Sources: Brook et al., Circulation 2010;121:2331–78, and Block and Calderón-Garcidueñas, Trends Neurosci 2009;32(9):506–16.

system, a special committee of the American Heart Association reviews the evidence that both acute and chronic particulate exposure lead to an imbalance between the sympathetic and parasympathetic portions of the autonomic nervous system [6]. This imbalance is thought to have multiple effects including vasoconstriction and an increase in blood pressure, an increased tendency for platelets to stick together and form clots, and several effects on cardiac rhythm. These include an increase in the baseline heart rate, a reduction in the normal variability of heart rhythm, and an overall increase in the probability for generating an abnormal heart rhythm, such as ventricular tachycardia or ventricular fibrillation, with consequences that are potentially fatal. Many of these same effects occur after particulates and/or their constituents including metals and organic compounds associated with particles enter the blood. Again, this is thought to lead to constriction of blood vessels, hypertension, and dysfunction of the cells that form the interface between the blood and the blood vessel (endothelial cells). This may contribute to the formation of atherosclerotic plaques or acceleration in the rate at which plaques form and possibly an increase in the tendency of platelets to stick together. The presence of atherosclerotic plaques increases the tendency of platelets to aggregate. Aggregations of platelets can form blood clots, and are major risk factors that lead to acute myocardial infarcts, transient ischemic attacks, and strokes.

Systemic oxidative stress that occurs in the lung has the potential to lead to systemic oxidative stress and inflammation in other organs or organ systems, as shown in figure 6.2. Again, the effects are multiple and varied. Inflammation and oxidative stress act on endothelial cells of blood vessels, accelerate or produce atherosclerosis, and increase the tendency for blood to clot and form thrombi. This is made worse by a simultaneous reduction in fibrinolysis, the normal tendency for blood clots to resolve spontaneously. Epidemiological and animal studies have gone as far as to suggest that these effects may mediate the development of insulin resistance and abnormal lipid metabolism [7–9]. This is dealt with more fully in chapter 11.

The epidemiological evidence for cardiovascular effects of fine particles over a period of days or less varies [6]. The best evidence links these particles to variations in the heart rate. The evidence for hypertension, systemic inflammation, activation of endothelial cells that line arteries, altered coagulation of the blood that affects blood vessel and endothelial cell function is moderate. Weaker evidence links short-term exposure to arrhythmias, cardiac ischemia, and systemic oxidative stress. Evidence linking exposure over months to years to cardiovascular events is weaker than the evidence for short-term effects [6]. These data link systemic inflammation, endothelial cell activation with abnormal blood coagulation, and heart-rate variability with longer exposures.

Figure 6.2 also illustrates how fine particles affect the brain by direct entry by crossing the cribiform plate, a thin bone that separates the nasal cavity from the brain. Its many small holes allow neurons mediating the sense of smell to gain access to inhaled air and to conduct fine particles into the brain. This route of entry is augmented by delivery of small particles and their constituents via the blood that flows to the brain. This triggers neural inflammation, activation of microglia, damage to lipids and DNA, and the proliferation of astrocytes, a marker of brain injury [12].

Suffice it to say that the cellular mechanisms that underlie the development of oxidative stress, inflammation and their consequences are extremely complex, as shown in figures 6.1 and 6.2. The details of these mechanisms are far beyond the scope of this book. The reviews by Brook et al. and Valko et al. are excellent sources of more information, including references to primary sources [4,6].

Reactive oxygen species, oxidative stress, and inflammation are thus all related and tied closely to coal-derived air pollutants. *Koyannis-quatsi*—the imbalances caused by pollutants—disrupt a large variety of processes that are essential for normal health.

The story does not end here. In the case of particulate matter, there is evidence that the very smallest particles cross the alveolar surfaces and enter the bloodstream directly. From there, they are distributed to all of the organs of the body where they have the potential to induce oxidative stress and inflammation. In addition there are cells in the alveoli called dendritic cells. They convey antigens, such as particulate matter, to other parts of the immune system and thus play a role in the development of immunological reactions to them.

Air Pollutants and the Central Nervous System

The brain is unique among the organs in that there is a barrier between it and the molecules, ions, and other constituents of the blood. This is the so-called blood–brain barrier (BBB)—without it the brain would not function. It protects the brain by excluding the biologically active ions, amino acids, and so forth, that are in the blood. Despite the presence of the BBB, there is evidence that particulates enter the brain directly [10,11]. Fine and ultrafine particles deposited in the nose appear to enter the neurons that mediate the sense of smell (olfactory nerve). From there they pass via these nerve cells to enter the limbic system of the brain. In addition to mediating the sense of smell, the limbic system mediates a variety of other functions, including emotions and, importantly, memory. Once in the brain, particulate matter has the potential to create ROS, cause inflammation, and activate microglia [12]. Microglia are cells in the brain that scavenge damaged cells and participate in immunological reactions. Neuroimaging techniques that employ tracers that bind selectively to microglia may make it possible to study the effects of pollutants in living individuals [12]. This has been done successfully in a variety of neurodegenerative disorders [13].

Evidence from in vitro studies of cultured neurons and glia has shown that particulates selectively damage neurons that employ dopamine as a neurotransmitter [14]. Loss of dopaminergic neurons is the hallmark of Parkinson's disease. Although this study used diesel exhaust particles, it is quite possible that particles from other sources cause similar, if not identical, responses. Investigators in Mexico have studied the brains of humans and dogs that lived in highly polluted Mexico

City and compared them to brains from nonpolluted sites [10,15]. The brains from animals and people that lived in regions of high pollution had evidence for increased activity of enzymes involved in inflammation and for increases in the concentration of the form of amyloid that is characteristic of Alzheimer's disease [10,15]. These authors conclude, "These findings suggest that exposure to severe air pollution is associated with brain inflammation and [amyloid] accu mulation, two causes of neuronal dysfunction that precede the appearance of neuritic plaques and neurofibrillary tangles, hallmarks of Alzheimer's disease."

Resistance to Oxidative Stress

The brain, and most if not all other organs, has a remarkable ability to heal itself and resist and recover from injury. The nervous system also has the capacity to rearrange itself and transfer function from an altered, damaged, or destroyed location to another undamaged or healthy portion. This phenomenon is referred to as neural plasticity. For example, in patients who suffer the loss of hearing in one ear, perhaps due to a tumor affecting the auditory nerve, other areas of the brain are recruited to respond to sounds [16]. These new sites are regions that do not respond to auditory stimuli in people who hear normally. Similar evidence for neural plasticity has been found after strokes and many other conditions in which the brain has been either damaged or subjected to altered sensory input.

Imagine an organ that is completely normal and unaffected by any adverse stimuli. Typically a sequence of low-level assaults has no effect, a situation equivalent to "normal" in figure 6.1. However, at some point this repeated assault begins to have an effect and damage begins. At this stage there may not be any loss of function that might be reported by a patient as a symptom, such as shortness of breath, or on a formal test of respiratory function. This is a window of golden opportunity—the opportunity to eliminate the stressor or the opportunity to make a therapeutic intervention. This is "tier 2" in the figure. If this opportunity is missed, the stress continues until it reaches the point where symptoms occur or laboratory tests become abnormal, or both. This is the point where a disease, such as asthma, is diagnosed. This stage is shown by "tier 3" in the figure. Continued damage eventually results in destruction of the organ or death. This is analogous to the multi-tiered response to oxidative stress shown in figure 6.1.

Ozone, and Oxides of Nitrogen and Sulfur

It is quite a lot easier to envision the mechanisms that make ozone, SO_x, and NO_x so toxic. They are irritating gases, something that is evident to anyone who inhales them. Therefore many of the steps between inhalation and the production of an inflammatory response and the generation of ROS are bypassed. Indeed, as explained in a contemporary review, they generate radicals directly in various tissues [4].

Mercury

As explained in chapter 3, coal plants are an important source of mercury that is introduced into the environment as the result of human activity. Like other metals, mercury can exist in many forms. Although coal-derived mercury enters the environment in an elemental form, it is converted to methylmercury by bacteria in waterways. Methylmercury is persistent, highly toxic, crosses cell membranes easily, and bioaccumulates in fish and animals that eat fish, including humans.

Although there are several well-studied tragic examples of mercury poisoning, the mechanisms by which methylmercury exerts effects are not well understood. The toxicity of methylmercury is thought to be related to its affinity and propensity to combine with molecules containing thiol groups [17]. Thiols are sulfur-containing analogues of alcohols. Alcohols contain molecules of oxygen bound to hydrogen, an OH group, whereas thiols contain molecules of sulfur bound to hydrogen, an SH group. Cysteine is an amino acid that possesses a thiol group. It is present in a large number of proteins that, in turn, form basic structural elements of cells. Thiol groups are also found in many of the subcellular components that mediate neurotransmission, the means by which neurons communicate with other neurons or target organs, such as muscles. When mercury interferes with cellular components that contain thiol groups, a variety of abnormalities may ensue. There is also evidence that methylmercury increases the rate of ROS formation, contributing to or initiating the multi-tiered response shown in figure 6.1

Animals fed methylmercury appear to develop apoptotic or necrotic cell death, in a concentration-dependent manner. Higher doses cause necrosis, possibly due to the effects of methylmercury on mitochondria, the subcellular organelles that are critically important in the production of energy. Lower doses produce effects that, on microscopical examination, are more consistent with apoptotic cell death.

Microtubules also appear to be highly sensitive to the deleterious effects of methylmercury [17]. These subcellular components are essential for a number of important processes, not the least of which includes cell proliferation. Microtubules are also required to transport various essential elements from one part of a cell to another. For example, in the case of neurons in the spinal cord, the cell body, the site of molecular assembly, may be several feet away from the site where the neuron connects to activate a muscle. Axonal transport is the process by which some molecules produced in the cell body are moved to synapses.

Neuronal migration also depends on the integrity of microtubules. Early in the process of the development of the brain, the cells destined to form the layers of the cortex originate far from their final destination. During the process of development, these neurons migrate from their original site of formation to their eventual destination. When migration is disrupted, neurons may be unable to move properly to their normal, final destination. The resulting developmental abnormality disrupts normal brain function and may cause epilepsy, intellectual disability, or both.

References

1. Bateson TF, Schwartz J. Children's response to air pollutants. J Toxicol Environ Health Part A 2008;71(3):238–43.

2. Trasande L, Thurston GD, Trasande L, Thurston GD. The role of air pollution in asthma and other pediatric morbidities. J Allergy Clin Immunol 2005;115(4): 689–99.

3. Nel A, Xia T, Madler L, Li N. Toxic potential of materials at the nanolevel. Science 2006;311(5761):622–7.

4. Valko M, Leibfritz D, Moncol J, Cronin MT, Mazur M, Telser J. Free radicals and antioxidants in normal physiological functions and human disease. Int J Biochem Cell Biol 2007;39(1):44–84.

5. Sies H. Oxidants and Antioxidants. London: Academic Press, 2011.

6. Brook RD, Rajagopalan S, Pope CA III, et al. Particulate matter air pollution and cardiovascular disease: an update to the scientific statement from the American Heart Association. Circulation 2010;121(21):2331–78.

7. Brook RD, Jerrett M, Brook JR, Bard RL, Finkelstein MM. The relationship between diabetes mellitus and traffic-related air pollution. JOEM 2008;50(1): 32–8.

8. Lockwood AH. Diabetes and air pollution. Diabetes Care 2002;25:1487–8.

9. Sun Q, Yue P, Deiuliis JA, et al. Ambient air pollution exaggerates adipose inflammation and insulin resistance in a mouse model of diet-induced obesity. Circulation 2009;119(4):538–46.

10. Calderón-Garcidueñas L, Maronpot RR, Torres-Jardon R, et al. DNA damage in nasal and brain tissues of canines exposed to air pollutants is associated with evidence of chronic brain inflammation and neurodegeneration. Toxicol Pathol 2003;31(5):524–38.

11. Oberdorster G, Sharp Z, Atudorei V, et al. Translocation of inhaled ultrafine particles to the brain. Inhal Toxicol 2004;16(6–7):437–45.

12. Block ML, Zecca L, Hong JS. Microglia-mediated neurotoxicity: uncovering the molecular mechanisms. Nat Rev Neurosci 2007;8(1):57–69.

13. Kannan S, Balakrishnan B, Muzik O, Romero R, Chugani D. Positron emission tomography imaging of neuroinflammation. J Child Neurol 2009;24(9): 1190–9.

14. Block ML, Wu X, Pei Z, et al. Nanometer size diesel exhaust particles are selectively toxic to dopaminergic neurons: the role of microglia, phagocytosis, and NADPH oxidase. FASEB J 2004;18(13):1618–20.

15. Calderón-Garcidueñas L, Reed W, Maronpot RR, et al. Brain inflammation and Alzheimer's-like pathology in individuals exposed to severe air pollution. Toxicol Pathol 2004;32(6):650–8.

16. Lockwood AH, Wack DS, Burkard RF, et al. The functional anatomy of gaze-evoked tinnitus and sustained lateral gaze. Neurology 2001;56(4):472–80.

17. Castoldi AF, Coccini T, Ceccatelli S, Manzo L. Neurotoxicity and molecular effects of methylmercury. Brain Res Bull 2001;55(2):197–203.

7

Basic Health Considerations

Health is a state of complete physical, mental and social well-being and not merely the absence of disease or infirmity.
—World Health Organization, 1948

This simple, elegant statement has stood the test of over sixty years of time. It recognizes the fact that even if you are not sick, in the traditional medical sense of suffering from a disease, you may not be healthy! This may seem paradoxical to many.

Beginning with this definition, it is possible to broaden our understanding of what determines health. The determinants begin with the individual and expand out to the society in which we individuals and groups of individuals live. We begin with ourselves—our personal behavioral characteristics, attitudes, and values. Moving from the individual to the group, the social and political environment in which we live is critical. Life in a home, family, or group, where warm, loving relationships exist is essential. Political or social oppression precludes a state of optimum health. Poverty is unhealthy. Expanding out to another sphere, the physical environment has its effect. Populations, groups of people, families, and individuals who are forced to live in environments where the physical conditions fail to meet appropriate standards cannot be considered to be healthy. It is this physical environment that is the major focus of this book.

The authors of the landmark series on energy and health, published in *The Lancet* in 2007, make important points about the relationship between energy and health [1]. The availability of energy has a large impact on physical conditions of the environment in which we live. Energy is needed to preserve and transport food, to keep us warm or cool, to produce the goods, services, and medicines that too many of us take for granted, to illuminate the night, and almost every other element of modern life. Without a relatively bountiful and inexpensive supply of energy, civilization as we know it would not exist. As many as 2.4 billion people may lack access to clean energy and live under conditions where they are exposed to dangerous levels of indoor pollution caused by inefficient burning of wood, dung, and other potentially renewable fuel [1]. Others, including too many who live in wealthy nations with bountiful supplies of energy, live in states of what the *Lancet* authors term "energy poverty," a state of energy deprivation. Too many of us who enjoy plentiful sources of energy that seem to be cheap on the surface, pay a price by exposure to the pollutants that are created by burning fossil fuels.

Yet another dimension needs to be added to the health equation: energy security. Arab oil embargoes, the shutoff of natural gas to Europe by Russia and to Chile by Argentina, and rising demands for fossil fuels by many nations are all evidence for international tensions that contribute to much of the political instability of our time.

When Thomas Edison introduced electricity as a commodity on that September afternoon in 1882, there is no way that he could have foreseen all of the consequences of throwing the switch that started the flow of energy to the nearby offices of J. Pierpont Morgan. But even then there were hints of what was to come. Although the lights glowed in Morgan's office, the streets were crowded by deliveries of coal and the removal of ash, and the air was sullied by the smoke from the boilers. Indeed there is a very close relationship between energy and health. Too few are aware of this and even fewer are prepared to take the steps necessary to promote health as it was defined by the United Nations from an energy perspective.

Lessons from Evidence-Based Medicine

Evidence-based medicine is, ". . . the conscientious, explicit, and judicious use of current best evidence in making decisions about the care of individual patients. The practice of evidence-based medicine means integrating individual clinical expertise with the best available external clinical evidence from systematic research" [2]. It is imperative to extend these principles beyond the individual patient to populations of individuals.

Bias is the enemy of good medical decision-making. In an attempt to control for bias, the American Academy of Neurology and similar organizations have developed rigorously defined grading systems to rate the strength of the evidence published in peer-reviewed studies, as shown in table 7.1 [3]. Although there are some studies that meet class I or II criteria that involve environmental pollutants, many are in the class III category. Class I and II studies are typically interventional in nature—the system is perturbed in some way, for example, by subjects inhaling ozone, and the effects are assessed and compared to the inhalation of a placebo. Some of the limitations are immediately obvious. Do ethical considerations permit the proposed study? How can investigators maintain blinding when a test compound such as ozone is noxious, and reactions of the participants (exclusive of the outcome measure) give important clues concerning the nature of the inhaled compound? Randomized trials may not be possible because of logistical and cost considerations. Alternatives are needed. Observational studies are one such alternative. These are systematic investigations of the world as it exists in the absence of any intervention by the researcher.

There is a hierarchy among observational studies. The case report is at the bottom of the ladder. The descriptions of the hapless chemistry professor who died after an accidental exposure to dimethylmercury, presented in chapter 10, is an excellent example of a good case report

Table 7.1
Classification scheme for weighting medical evidence

Class	Characteristics of the class
I	Randomized controlled trials with appropriate blinding, masking, and randomization procedures in a representative population with defined outcome measures, inclusion/exclusion criteria, and other characteristics.
II	Randomized controlled trials not fully meeting one criterion for class I study.
III	All other controlled trials including those with natural history controls where outcome is independently assessed or derived by objective outcome measurement. Many epidemiological studies fall into this class.
IV	Studies not meeting class I, II, or III criteria, including consensus opinions and expert opinions.

[4]. It illustrates a point vividly but does not make a huge contribution to our understanding of the public health consequences of mercury in the environment because it constitutes but a single example of a phenomenon. Case reports typically do not provide evidence of causality. A case series is one rung up. At least there is more than the one example of the phenomenon. However, the series fails to test a hypothesis concerning causality. Ecological studies are next. These are observational studies in which a population rather than an individual forms the unit that is being investigated. My report of the highly significant association between the prevalence of type II diabetes mellitus and the number of pounds of toxicants released into the air reported in the Toxics Release Inventory is an ecological study [5]. Ecological studies may seem compelling, but the absence of hypothesis-testing does not allow the investigator to make rigorous cause-and-effect conclusions.

Analytical observational studies are capable of testing a hypothesis and take the form of case control or cohort studies. Case control studies are efficient ways to investigate rare phenomena and usually require smaller samples than cohort studies. Case control studies begin by identifying affected individuals, or cases, and retrospectively determining which among these was exposed to an agent, such as a toxicant, that may be the cause of the condition. The cases are compared to a control population. These controls are selected or matched with cases so that demographic criteria are similar to the individuals among the cases, in accord with age, sex, race, and so forth. In the next step it must be determined how many of the controls have been exposed. By comparing the probability of the exposure in the cases and controls, it is possible to determine whether the exposure is more common in the cases than the controls. It is common to identify two or three times as many controls as cases to increase the statistical power of the study. Biases can creep into the selection of both cases and controls.

In cohort studies, individuals who meet specific inclusion criteria are identified and classified according to whether or not they have been exposed. They are then followed for a period of time during which examinations determine whether or not they develop the disease in question. The Women's Health Initiative study provides an excellent example [6]. Within the cohort enrolled at the beginning of the study, a subset of postmenopausal women was identified. Exposures to particulate matter were linked to each individual using data from EPA monitoring sites. Enrollees were then followed to determine whether predetermined endpoints, such as death or the development of cardiovascular disease or a cerebrovascular event occurred [6]. All three of these endpoints were significantly associated with increases in particulate matter exposures. In cohort studies, the direction of inquiry moves forward in time, whereas in case control studies, the inquiry is retrospective.

Significant correlations may suggest, but do not prove, cause-and-effect relationships. Nevertheless, cause-and-effect relationships, the most desirable of conclusions, become more likely as various criteria are met. The strength of the association is perhaps the most important of these criteria. Highly significant associations are much more likely to indicate a causal relationship than ones that just barely meet criteria for statistical significance. Consistency with other studies is important. One study alone carries less weight than ten that all show the same association. This is why replication of results is often of critical importance and forms the rationale for funding similar studies in different countries.

Other factors in addition to correlation are important in determining whether a cause-and-effect relationship is present. For example, if the relationship between an exposure and a disease is plausible, a causal relationship is more likely. It makes sense that inhaling ozone, an irritating gas, affects pulmonary function and may precipitate an acute attack of asthma. Temporal relationships are also critical. If hospital admissions for myocardial infarcts or strokes follow immediately after a transient peak in the concentration of a pollutant in the air, the cause-and-effect relationship is more likely. Other factors that should be considered are whether there is a dose–response effect. Does an endpoint become more likely if the concentration of a pollutant increases? Is there evidence from animal experimentation to support a causal relationship? Have alternative explanations been explored within the data set or by other studies? Finally, if the exposure is halted or reduced significantly, does the endpoint become less probable? For many of the large observational studies most if not all of these criteria are met. Thus a cause-and-effect relationship becomes extremely likely.

The result of observational studies is commonly expressed as a relative risk. The relative risk is the ratio of the probability of developing a condition in the exposed versus the nonexposed population and is commonly associated with a confidence interval, usually set at 95%. The confidence interval is somewhat difficult to explain rigorously; however, it serves as a measure of the amount of random error that is present. For example, if the 95% confidence interval is narrow, such as between 1.45 and 1.55, it means that if the experiment were done 100 times, the result would lie between these two numbers in 95 of those trials. If the confidence interval is large, then there is a correspondingly large amount of random error in the comparison. In the following chapters, I will use the acronym CI in parentheses to indicate the confidence interval. Hopefully this will allow readers with a background in statistics to make additional inferences without being a source of excessive confusion to others.

The odds ratio is another statistic used to evaluate the risk or probability of an outcome if an exposure is present. It is an indicator of how much more likely an exposed individual is to experience an outcome, such as stroke or myocardial infarct, compared to one who is not exposed.

I am often approached by individuals who are absolutely convinced that their medical problem is the direct result of their exposure to a specific pollutant. I hear: "My cancer was caused by the coal ash spread on the road in front of my house" or some similar statement all of the time. This poses a real problem. On the basis of the epidemiological data, this scenario might be true, but it is nearly impossible to prove the point. Litigators know this very well. When dealing with large numbers of individuals affected by a common condition, such as a myocardial infarct, it is almost always impossible to determine whether specific exposure to coal-derived pollutants caused an infarct on that given day, but did not cause the infarct affecting the patient in the next bed. It is necessary to understand probability and relative risk, namely the results of observational studies, to deal with this conundrum. These studies tell us, with a high level of confidence expressed in statistical terms, that the risk is increased by some defined amount as the consequence of a given exposure. Difficult as this may be to understand and apply to a given individual, these are the data that the EPA uses to estimate the effects of pollutants and, importantly, the benefits to be gained by reducing the concentration of coal-derived pollutants in the air. By extrapolating the results of studies of a given study population to everyone, it is possible to quantify health benefits and risks. This is how the EPA made its estimate that the amended Clean Air Act will save 230,000 lives per year by 2020 (see chapter 13). It is impossible to tell an individual that "You were spared," but it is possible to tell everyone in the country that 230,000 will be spared.

Effects of Pollutants on General Health

Infant Mortality and the Neonatal Period

There are significant effects of air pollutants on the respiratory health of children. These are dealt with in chapter 8, in the discussion of diseases of the respiratory system. The effects on other more general measures of the health of children are much less clear.

In an attempt to clarify the effects of air pollution on children's health, Glinianaia et al. reviewed relevant literature published between the beginning of 1996 through the end of 2003 [7]. They focused on particulate matter. The variety of methodological approaches, corrections for potential confounding factors, geographical areas, and so forth, made it somewhat difficult to draw general conclusions. After a careful review of eight studies they found little consistency between particulate matter and total infant mortality, defined as the total number of deaths during the first year of life for each 1,000 live births. While some studies indicated a positive association, inconsistences, the failure of some to achieve statistical significance, and absences of correlations confounded their attempts to draw firm conclusions.

Similar limitations were encountered in a review of deaths during the neonatal period, which includes days 0 to 27 of life [7]. Five studies from different geographical regions failed to reveal any association between particulates and nonrespiratory deaths during this period.

During the postneonatal period, extending from day 28 of life to the end of the first year, the picture was clearer [7]. Five of the studies they reviewed reported consistent positive associations between particulates and deaths during that period.

A subsequent review of thirteen studies of air pollution in Canada did not clarify these issues any further [8]. It is possible that some of the inconsistences among the data summarized in these two reviews will be eliminated as more consistent time-series data for pollutant concentrations and public health records become available.

A more highly focused study of pregnancy outcomes in Vancouver, British Columbia, took advantage of the daily pollutant level measuring systems and health data available for this large metropolitan region [9]. They found an association between low birth weight (less than 2,500 grams) and the mother's exposure to sulfur dioxide during the first month of pregnancy (odds ratio = 1.11, 95% CI = 1.01–1.22 for a 5 ppb increase). Preterm birth, a live birth that occurred at less than 37 full weeks of gestation, was also associated with sulfur dioxide exposure (odds ratio = 1.09, 95% CI = 1.01–1.19 for this same increase). Intrauterine growth retardation, a birth weight at less than the 10th percentile for the gestational week and sex of the baby, was also correlated with exposure to oxides of sulfur and nitrogen (odds ratios were 11.07, 95% CI 1.01–1.13 and 1.05 95% CI = 1.01–1.10 for 5 ppb increases, respectively). Thus these two coal-derived pollutants appear to have consistent adverse effects on pregnancies.

Reports describing environmental effects on children between 1 and 18 focus primarily on the respiratory system. However, exposures that occur during childhood may not manifest themselves as disease until much later in life. This point was made by Goldman almost two decades ago [10]. In that paper she described the explosion in the number of chemicals introduced into the environment with little understanding of their potential health effects. Although the examples she uses to illustrate her point are largely unrelated to coal-derived toxicants, the principles of multiple exposures, toxin synergy, and delayed effects apply. The focus returns more clearly to coal in a report by Millman et al. [11]. Their study deals with air pollution threats to children in China, where between 70% and 75% of its energy needs come from burning coal, leading to the release of large amounts of particulate matter, polycyclic aromatic hydrocarbons, mercury, and other metals. The risk in this population is increased by the fact that many Chinese burn coal indoors to cook their food. Cancers and other chronic respiratory diseases, such as emphysema and chronic bronchitis, are characterized by long periods of exposure before a disease appears. The authors conclude that China's future is threatened by the pollutants discharged during this period of exuberant economic growth.

Deaths among Adults

Many of the largest epidemiological studies of the health effects of air pollutants refer to changes in life expectancy, morbidity, or mortality. They do not fit neatly into the chapters dealing with diseases of specific organ systems and are included in this chapter.

Based on a variety of studies that linked particulate matter exposure to overall mortality, Pope and others hypothesized that reductions in the concentration of particulate matter would be associated with increases in longevity. They utilized measured or calculated concentrations of particles 2.5 μm (microns) or less in size with data that were available for 51 metropolitan areas in the US dating from the early 1980s and the late 1990s. Life expectancy data for the 215 county units that made up these 51 metropolitan areas were retrieved from national mortality and census statistics. Changes in life expectancy were correlated with changes in particulate concentration data for each of these areas and for the aggregated group. Controls for socioeconomic status, demographic variables, and smoking were applied. A decrease in the concentration of 10 μg (micrograms) per cubic meter was associated with an increase in life expectancy of 0.61 ± 0.20 years (plus or minus the standard error, $p = 0.004$). Some cities were big winners. In the Buffalo, New York, metropolitan area, the small particle concentration fell by 13 μg per cubic meter and the life expectancy rose by 3.4 years. Reductions in the concentration of particulates were thought to account for approximately 15% of the improvements in longevity in these regions.

In the much earlier study, usually referred to as the Harvard Six Cities Study, the investigators prospectively followed over 8,000 white inhabitants of six different cities in the United States. These cities included Watertown, Massachusetts; Harriman, Tennesee, located just a few miles from the Kingston Fossil Plant, site of the 2008 coal ash slurry spill (see also chapter 4 for additional details about this facility); Steubenville, Ohio; portions of St. Louis, Missouri; Portage, Wisconsin, including Wyocena and Pardeeville; and, Topeka, Kansas. Steubenville is located within 50 kilometers of five large coal-fired boilers and within 100 kilometers of 17 more additional boilers [12]. Enrollment inclusion criteria restricted individuals to those between 25 and 74 years of age who underwent tests of lung function and completed a standard questionnaire that included information about sex, weight, height, educational level, smoking and occupational histories, and a medical history. Follow-up lasted for 14 to 16 years and included 111,076 person-years. Death certificates were obtained for 98% of the 1,430 participants who died. The quality of outdoor air was measured at a site located in each city and included total suspended particles (small particles as well as those with other sizes), sulfur dioxide, ozone, and suspended sulfates. The levels of most pollutants were the highest in Steubenville and the lowest in Topeka. Although smoking had the greatest effect on mortality, the investigators "observed statistically significant and robust associations between air pollution and mortality." The residents of Steubenville were 1.26 times as likely to die from pollutants than residents of Topeka (95% CI = 1.08–1.47) after adjustments were made to account for the effects of smoking, body mass index, and an education level less than completion of high school. By confining this study to white Americans, the investigators may have underestimated the number of deaths, as omitted populations, for example, blacks, have a much higher prevalence of cardiovascular disease than whites [13]. Extending this study for an additional six years yielded similar results, with a rate ratio of 1.16 (95% CI = 1.07–1.26).

The American Cancer Society Cancer Prevention Study II was similar, but included more participants from more locations [14]. A total of 552,138 individuals were linked to air pollution data from 151 metropolitan areas in 1982. Mortality information was determined through the end of 1989. Controls for smoking, education, and other risk factors were employed. The relative risk ratio for all-cause mortality in the most compared to the least sulfate polluted area was 1.15 (95% CI = 1.09–1.22) and somewhat higher for fine particulates where the risk ratio was 1.17 (95% CI = 1.09–1.26). In a follow-up analysis that extended the observation period to the end of 1998, the all-cause mortality rate was found to increase by approximately 4% for an increase in fine particle concentration of 10 $\mu g/m^3$ (micrograms per cubic meter) [15].

Eftim et al. compared the results of both the Harvard Six Cities and the American Cancer Society data to data from Medicare subscribers [16]. They found that a 10 μg per cubic meter increase in the small particle concentration was associated with an increase in all-cause deaths of 10.9% (95% CI = 9.0–12.8) for the counties represented in the American Cancer Society study and an increase of 20.8% (95% CI = 14.8–27.1) for this in counties included in the Harvard Six Cities study. These landmark studies clearly link exposure to coal-derived air pollutants to increases in deaths from all causes. The ensuing chapters will examine this relationship in greater detail for disease of the respiratory, cardiovascular, and nervous systems.

The Bottom Line

The evidence reviewed in this chapter shows significant correlations between exposure to various air pollutants produced by burning coal and general measures of morbidity and mortality. These effects were observed at all ages and in multiple countries. The good news is that reductions in pollutants, particularly particulate matter, are associated with improved survival [14]. As discussed in detail in chapter 13, these data should provide an evidence-based argument for improving air quality.

References

1. Wilkinson P, Smith KR, Joffe M, Haines A. A global perspective on energy: health effects and injustices. Lancet 2007;370(9591):965–78.

2. Sackett DL, Rosenberg WM, Gray JA, Haynes RB, Richardson WS. Evidence based medicine: what it is and what it isn't. BMJ 1996;312(7023):71–2.

3. French J, Gronseth G. Lost in a jungle of evidence: we need a compass. Neurology 2008;71(20):1634–8.

4. Nierenberg DW, Nordgren RE, Chang MB, et al. Delayed cerebellar disease and death after accidental exposure to dimethylmercury. N Engl J Med 1998; 338(23):1672–6.

5. Lockwood AH. Diabetes and air pollution. Diabetes Care 2002;25:1487–8.

6. Miller KA, Siscovick DS, Sheppard L, et al. Long-term exposure to air pollution and incidence of cardiovascular events in women. N Engl J Med 2007;356(5):447–58.

7. Glinianaia SV, Rankin J, Bell R, Pless-Mulloli T, Howel D. Does particulate air pollution contribute to infant death? A systematic review. Environ Health Perspect 2004;112(14):1365–71.

8. Koranteng S, Vargas AR, Buka I. Ambient air pollution and children's health: a systematic review of Canadian epidemiological studies. Paediatr Child Health 2007;12(3):225–33.

9. Liu S, Krewski D, Shi Y, Chen Y, Burnett RT. Association between gaseous ambient air pollutants and adverse pregnancy outcomes in Vancouver, Canada. Environ Health Perspect 2003;111(14):1773–8.

10. Goldman LR. Children—unique and vulnerable. Environmental risks facing children and recommendations for response. Environ Health Perspect 1995;103 (suppl 6):13–18.

11. Millman A, Tang D, Perera FP. Air pollution threatens the health of children in China. Pediatrics 2008;122(3):620–8.

12. Keeler GJ, Landis MS, Norris GA, Christianson EM, Dvonch JT. Sources of mercury wet deposition in eastern Ohio, USA. Environ Sci Technol 2006;40(19): 5874–81.

13. Roger VL, Go AS, Lloyd-Jones DM, et al. Heart disease and stroke statistics—2011 update: a report from the American Heart Association. Circulation 2011;123(4):e18–209.

14. Pope CA III, Thun MJ, Namboodiri MM, et al. Particulate air pollution as a predictor of mortality in a prospective study of U.S. adults. Am J Respir Crit Care Med 1995;151(3 pt 1):669–74.

15. Pope CA III, Burnett RT, Thun MJ, et al. Lung cancer, cardiopulmonary mortality, and long-term exposure to fine particulate air pollution. JAMA 2002; 287(9):1132–41.

16. Eftim SE, Samet JM, Janes H, McDermott A, Dominici F. Fine particulate matter and mortality: a comparison of the six cities and American Cancer Society cohorts with a medicare cohort. Epidemiology 2008;19(2):209–16.

8

Diseases of the Respiratory System

It is almost intuitively obvious that air pollutants will have important effects on respiratory health. Virtually all airborne pollutants gain access to the body via the respiratory tract. Thus it is no surprise that this important system is affected significantly by pollutants discharged into the atmosphere by electrical utilities that burn coal. These effects include de novo production of a condition that did not exist prior to an exposure, such as asthma, chronic bronchitis, emphysema, or cancer and/or exacerbation of a previously existing illness. Pollutants commonly trigger exacerbations of asthma, as well as of chronic bronchitis and emphysema, which usually occur together and are referred to as chronic obstructive pulmonary disease.

The most recent mortality data from the Centers for Disease Control list chronic lower respiratory diseases, such as asthma, emphysema, and chronic bronchitis, as the third ranked cause of death among Americans in 2008, the most recent year for which these data are available [1]. This was a 7.8% increase from the prior year. The Centers recommend caution in the interpretation of this increase as the methodology for computing this data point changed, effective with the current report. Methodological issues aside, the 141,075 deaths reported in 2008 are not to be trivialized.

Asthma

Epidemiology and Demographics of Asthma
Asthma is an excellent example of a pollution-related disease that has a disproportionate effect on children and other sensitive populations, as

shown in table 8.1 [2]. In addition there are marked differences among ethnic groups and stratification of risk by income levels. As shown in the table, children are more likely to have asthma than adults. Among the children, boys are more likely to be affected than girls, a gender ratio that is reversed in adults. Among the children, those of Puerto Rican descent were the most likely to have asthma and white children were the least likely to be asthmatic. Asthma was most prevalent among the poor for all ages and least prevalent among those living at twice the federal poverty level or higher. In other words, asthma is most likely to affect individuals who are least able to speak up in their own defense.

The prevalence of asthma in children has changed substantially during the past three decades. The prevalence of 3.6% in 1980 rose to a peak of 7.5% in 1995 [3]. During that period the asthma prevalence was determined by the answer to the question, "During the past 12 months, did anyone in the household have asthma?" In 1997, the survey questionnaire was redesigned to include two new questions, "During the past 12 months did [child's name] have an asthma attack?" and "Has a doctor

Table 8.1
Current asthma prevalence

Characteristic	Children (%)	Adults (%)	Total (%)
White, non-Hispanic	8.2	7.7	7.8
Black, non-Hispanic	14.6	7.8	9.5
Multiracial	13.6	15.1	14.8
Hispanic of Puerto Rican descent	18.4	12.8	14.2
Male	10.7	5.5	6.9
Female	7.8	8.9	8.6
Poor, below federal poverty level	11.7	11.0	11.2
Below twice federal poverty level but not poor	9.9	7.9	8.4
Equal to or above twice federal poverty level	8.2	6.6	7.0
Total	9.3	7.3	7.8

Source: From National Health Interview Survey, 2006–2008 [2].
Note: Includes persons who answered yes to the questions, "Have you ever been told by a doctor or other health professional that you had asthma?" and "Do you still have asthma?" Children were ages 0 to 17 years, inclusive; adults were ages 18 years and older.

or health professional ever told you that [child's name] had asthma?" A third follow up question was added in 2001, "Does [child's name] still have asthma?" During the post-1997 period, the percent of children who had ever had asthma rose from about 11% to about 13.1% while the current asthma prevalence and asthma attack prevalence remained relatively constant. The data from the 2007 National Health Information Survey showed that the prevalence of asthma remained at "historically high levels" [3]. Those data showed that 9.1% of all children, some 6.7 million, currently had asthma and that 5.2% of children, or 3.8 million, or 60% of children who currently had asthma had one or more attacks during the past year. The most encouraging information from the survey showed that the number of patient visits to a physician's office or hospital outpatient facilities appears to have fallen from 2004 highs.

There are substantial differences in the prevalence of asthma among the residents of the different states, as revealed in the 2009 Behavioral Risk Factor Surveillance System [4]. The self-reported prevalence of asthma in adults ranged from a low of 9.6% in Minnesota to a high of 16.9% in Hawaii with a mean for the entire country of 13.4% among the 423,457 individuals surveyed [4].

Clinical Aspects of Asthma

Asthma is a chronic disease of the lungs characterized by inflammation and narrowing of the airways. Patients with asthma experience recurrent episodes of shortness of breath (dyspnea), a sensation of tightness in the chest, wheezing, and coughing. These symptoms are typically worst at night or early in the morning. Airway inflammation in asthmatics causes swelling that narrows a bronchial tree that has been previously sensitized to inhaled irritants, including many air pollutants. Exposure to an inhaled irritant causes further narrowing of the airways and the production of mucus that makes airways even narrower. Air moving through the narrowed airways causes the wheezing that can be heard with a stethoscope or by those next to the asthmatic patient or both. Narrowing of the airways also increases the amount of work that needs to be done by the diaphragm and the other muscles that move air in and out of the lungs. Under extreme conditions, asthmatics may become completely unable to breathe on their own and will die unless they receive emergency medical attention and mechanical ventilation.

During severe attacks the amount of air reaching the lungs may not be sufficient, and the lungs may fail to perform their task of exchanging carbon dioxide, produced by metabolic processes in the body, for life-giving oxygen. This can lead to a reduction of the amount of oxygen in the blood (hypoxia) and a high blood carbon dioxide level (hypercarbia). These may cause acidification of the blood by the accumulation of carbon dioxide that may, along with hypoxia, cause cardiac arrhythmias and death. Severe hypoxia affects normal brain function producing delirium or, in extreme cases, coma or death.

Acute attacks of asthma are often precipitated by triggers that cause an immune response to certain environmental stimuli. Asthmatics are more sensitive to these triggers than non-asthmatics, a condition known as hypersensitivity. There are many triggers, including dust, smoke, pollen, volatile organic compounds, particulate matter, ozone, and others. A detailed description of the immunological basis for attacks of asthma and the role of oxidative stress is far beyond the scope of this review but is depicted, in part, in chapter 6. Some of the pollutants discharged by coal-fired power plants or their products, such as the ozone formed from NO_x, and particulate matter may act as triggers and produce an asthmatic attack.

Global warming is likely to increase the concentration of airborne pollen from some plants, such as ragweed, and thereby contribute to the development of additional asthmatic attacks. For more information on health effects of global warming, see chapter 12.

Genetic variability accounts for some of the differences in the sensitivity of individuals to asthma triggers [5]. Genetic studies have identified differences in the susceptibility to ozone that are due to subtle differences in the structure of genes that mediate responses to oxidative stress and the inflammatory response. These differences in the structure of genes are known as polymorphisms. Thus the probability that an individual will develop asthma depends on exposure to a trigger and the individual's susceptibility to that trigger. There is a complex combination and interaction between genetic and environmental factors. For a review of the genetic susceptibility to the effects of air pollutants, such as ozone, particulates, nitrogen dioxide, and sulfur dioxide on respiratory function, see [5].

Children appear to be more susceptible to the development of pollution-related asthma attacks than adults. There are several explanations for this observed increase. According to a 2008 review, the susceptibility of children to the effects of air pollution is multifactorial and includes the following: (1) Children have different patterns of breathing than adults. (2) They are predominantly mouth-breathers, thereby bypassing the filtering effects of the nasal passages. This allows pollutants to travel deeper into the lungs. (3) They have larger lung surface areas per unit weight than adults. (4) They spend more time out of doors, particularly in the afternoons and during the summer months when ozone and other pollutant levels are the highest. (5) Children have higher ventilation rates than adults. (6) When active, children may ignore early symptoms of an asthma exacerbation and fail to seek treatment leading to attacks of increased severity [6]. There are other factors that are important. The diameter of the airways in children is smaller than in adults and therefore airways may be more susceptible to the effects of the airway narrowing that is characteristic of asthmatic attacks. These factors, combined with the possible adverse impact of pollutants on lung development and the immaturity of enzyme and immune systems that detoxify pollutants, may all contribute to an increase in the sensitivity of children to pollutants produced by burning coal [7].

Ozone, Air Pollution, and Asthma

Ozone is a highly reactive gas that consists of three molecules of oxygen. Ozone is formed by the reaction of volatile organic compounds with oxides of nitrogen in the presence of sunlight. Oxides of nitrogen are formed at the high temperatures in coal-fired boilers. Ozone is a powerful oxidizing agent that irritates the lungs at concentrations typically encountered in urban settings, particularly in summer months. For additional details, see chapter 3.

As of this writing, the primary National Ambient Air Quality Standard for ozone is 0.075 ppm (parts per million). Primary standards are those designed to protect public health. This standard was established after a controversial 2008 review. At that time the EPA administrator ignored the unanimous advice of the Clean Air Scientific Advisory Committee as reflected in the letter from Chairman Rogene F. Henderson to EPA Administrator Stephen L. Johnson, dated April 7, 2008, in which a primary 8 hour standard between 0.06 and 0.07 ppm was strongly recommended. After a lengthy review that included a proposed new standard and extensive period of public comment, President Barak Obama unexpectedly instructed the EPA to stop the ozone rule-making process and allow the 0.075 standard to remain in effect. This decision was seen widely as a concession to Republicans in the House of Representatives and angered many health professionals and environmentalists.

In the rule as it was proposed, the Agency appeared to rely on two important studies by Adams published in 2002 and 2006 [8,9], a study by Brown et al. published in 2008 [10], and a 2009 study by Schelegel et al. [11]. In the 2006 Adams study, the investigators obtained several standard measures of the lung's ability to function properly. These included a measurement of the maximum amount of air that can be exhaled in one second. This is known as the forced expiratory volume or $FEV_{1.0}$. They also measured the largest amount of air that can be exhaled after inhaling as much air as possible. This is known as the forced vital capacity (FVC). Finally they rated symptoms using a questionnaire known as the Total Symptom Scale during a controlled exposure to ozone. During the experiment the ozone concentration was increased from a baseline to a peak and than back again to the baseline over a 6.6 hour period. Compared to initial scores, there were more symptoms after 5.6 and 6.6 hours of exposure in the 0.06 part per million inhalation protocol. In addition the ratio between the $FEV_{1.0}$ and FVC for the 0.06 ppm exposure was significantly greater than it was when the participants were breathing filtered air. Thus the exposure to 0.06 ppm of ozone, the lowest proposed for the rule, resulted in statistically significant differences in objective tests of pulmonary function and subjective measures of symptoms. In the re-analysis of the Adams data obtained during steady-state exposures to this concentration, Brown et al. reported that 24 of the 30 subjects experienced a highly significant 2.85% reduction in the $FEV_{1.0}$ ($p < 0.002$). The $FEV_{1.0}$ fell by more than 10% in two of these subjects. In another study of exposure to 0.070 ppm of ozone, the investigators found a significant reduction in the $FEV_{1.0}$ (p values ranged from 0.01 to 0.0016, depending on the statistical model employed) [11]. They also found a significant increase in the symptom score.

As in many other investigations of normal individuals, these researchers recruited normal, healthy, young adult volunteers who had not had large ozone exposures in the past. This population is the least likely to suffer the consequences of ozone exposure. Children and adults in older age groups are the most likely to be affected by ozone. Even among those highly selected subjects, symptoms and objective changes in pulmonary function were found at the proposed 0.060 part per million 8 hour standard. It is quite possible that a genetically determined susceptibility to the effects of ozone was responsible for this substantial decrement in pulmonary function. Susceptible populations should be protected by the proposed standard. In fact they deserve more protection than more normal individuals.

In its provisional assessment of the health effects of ozone exposure, the EPA reviewed much of the pertinent literature [12]. These supplemented data are contained in the Agency's earlier report published in 2006 [13]. In the 2006 report, the EPA concluded that "the evidence supports a causal relationship between acute ambient O_3 exposures and increased respiratory morbidity outcomes resulting in increased [emergency department] visits and hospitalizations. . . ."

There are many studies linking increases in ozone to asthma and other pulmonary diseases (see Trasande and Thurston for a review [7]). One of the most compelling of these studies was published in 2003 [14]. The authors conducted a prospective cohort study of 271 children younger than 12 who had physician-diagnosed asthma. The children were divided almost equally into groups who did or did not use daily maintenance medications. The analysis examined the relationship between ozone levels below EPA standards, respiratory symptoms, and the use of rescue medications charted by the children's mothers on daily calendars. Rescue medications are needed to treat symptoms of asthma that occur in patients who regularly take medications designed to prevent asthma attacks. They found a significant association between ozone levels and symptoms as well as the use of rescue medications in the children who used daily maintenance medications. No significant relationships were found between ozone levels and symptoms or medication use in the children who did not take daily maintenance medications. Thus it appears that the threat to children posed by ozone is greatest among those with severe asthma, even when ozone levels are below the EPA standard.

Peel et al. studied the relationship between a one standard deviation increase in ambient air pollutant levels and emergency room visits for various respiratory problems, including asthma [15]. They found the strongest association between increases in 24 hour concentration of 10 micron particle levels, 24 hour increases in larger particles (10–100 nm [nanometers] in size), and 1 hour nitrogen dioxide concentrations with asthma attacks that occurred 6 to 8 days after the peak. During warm months there was a 2.6% increase in asthma admissions after a 0.025 ppm increase in the ozone concentration. There were shorter delays between peaks in the concentration of the 10 μm (micron) particles, ozone, nitrogen dioxide, and carbon monoxide and emergency room visits for upper respiratory infections.

The evidence linking levels of ozone and its nitrogen dioxide precursor to the development of new cases of asthma is less compelling than that linking ozone to asthma exacerbations. Gilmour et al. reviewed five studies that evaluated the potential effects of pollutants on the incidence of asthma [16]. A Dutch study of over 4,000 children enrolled at birth and followed for two years focused on nitrogen dioxide and small particles they attributed to traffic [17]. They found small but statistically significant associations between symptoms and the pollutants. Although this study focused on traffic as the source of the pollutants, burning coal must be considered as a possible source of these two pollutants. Similar results were found as the result of a second study of children in that age group in the Netherlands, Germany, and Sweden [18]. The Children's Health Study of more than 6,000 children from southern California evaluated a wide concentration range of ozone, particulates, oxides of nitrogen, and acids [19]. A significant association between ozone and asthma was confined to those children who participated in three or more sports. This result may be the consequence of the increases in breathing that are associated with exercise and the consequent increases in exposure to pollutants. Gilmour et al. conclude that the results of all five of the studies they reviewed "support a modest increase in the risk for air pollution in relation to phenotypes (observable characteristics or traits) relevant to asthma" [16].

The increase in susceptibility to pollutants appears to translate into pollution-related increases in infant mortality. Ritz et al. reported increases in the risk of death from respiratory causes, including sudden infant death, with rises in the concentration of carbon monoxide, 10 μm particles, and nitrogen dioxide [20]. Bateson and Schwartz also cite a study reporting between 4 and 7 fewer infant deaths per 100,000 live births with a reduction of total suspended particles of 1 µg/m³ (microgram per cubic meter) [6].

Chronic Obstructive Pulmonary Disease (COPD): Chronic Bronchitis and Emphysema

Epidemiology and Demographics of Chronic Obstructive Pulmonary Disease

Patients with COPD are included among those classified as having chronic lower respiratory diseases, currently ranked third as a cause of death in Americans [1]. About 5% of all deaths in the United States are due to COPD. Between 2000 and 2005, 718,077 deaths due to COPD were reported among Americans 25 years of age or older. During this interval there was an increase of 5% among men and an alarming 11% increase among women. Death rates were the highest for white men and lowest among black women. The reported rates per 100,000 Americans were: 80.2 for white men, 63.8 for black men, 60.3 for white women, and 29.9 for black women. As is the case for asthma, death rates varied substantially among states ranging from a low of 27.1 per 100,000 in Hawaii to a high of 93.6 per 100,000 in Oklahoma. About 75% of deaths due to COPD are attributed to smoking. Air pollution, including pollutants produced by burning coal, is responsible for many of the others.

Basic Facts

COPD is a heterogeneous group of diseases characterized by narrowing of the airway passages and an obstruction of air flow in the lungs that impedes normal breathing. Unlike asthma, these changes are permanent rather than reversible. Exposures to pollutants that produce oxidative stress and stimulate the immune system, including SO_x, NO_x, along with primary and secondary particulates, are important in the pathogenesis of COPD. Burning coal is an important source of these pollutants. The response in larger airways is referred to as chronic bronchitis. A productive cough (i.e., one that produces sputum) is characteristic of chronic bronchitis. Chronic bronchitis and emphysema commonly exist in the same individual. Emphysema is caused by the destruction of the alveoli in the lungs. Bullae, also called blebs, may form in the lungs as the alveoli are destroyed. Bullae, from the Latin bulla, meaning bubble, are blister-like cavities in the lung. Exacerbations of COPD may be triggered by an exposure to pollutants or by infections.

Pulmonary Inflammation and Air Pollutants

Inflammation of pulmonary tissues is a critical element in the pathophysiology of illness caused by air pollution. Reactive oxygen species appear to be central to this process. Several investigators have studied the response to particulates in experimental animals. Although they don't fully reproduce the human condition, animal models allow scientists to produce experimental conditions that are impractical or unethical to produce in human research participants. Roberts et al. instilled particles into the lungs of rats treated with a compound (dimethylthiourea) that is believed to blunt the response to reactive oxygen species [21]. After treatment the lungs of the animals were lavaged (rinsed with saline) and chemicals that are markers for damage to the lungs were measured. Treated animals had less evidence for damage to the lungs, less activity of genes that control the expression of proteins involved in cellular signaling (cytokines), and less evidence for inflammation of the lungs and other markers of pulmonary injury. In a subsequent study Rhoden et al. instilled standardized urban air particles (active agent) or saline (placebo control) into the lungs of rats [22]. Half of the animals in each group were treated with an inhibitor of superoxide anion formation (an anion involved in the production of oxidative stress). Pretreatment with the inhibitor blocked the deleterious effects of the particles, as shown by reductions in several markers of pulmonary inflammation.

These two studies are representative of many that have been performed using a variety of agents and techniques. Although performed in animals, they provide coherence (i.e., consistency) with a larger body of scientific evidence, that helps establish a cause-and-effect relationship between particulates and pulmonary disease. As noted above, inflammation is a critical element in the pathogenesis of attacks of asthma and exacerbations of COPD. It matters little whether the inflammation is caused by particulates or other pollutants.

Additional evidence to support the hypothesis that air pollutants produce oxidative stress is derived from many other studies. Recently Fitzpatrick et al. studied 65 children with severe asthma. Their study included 35 children whose forced expiratory volume was less than 80% of that predicted by age [23]. Bronchoalveolar lavage (rinsing the airway with saline) was performed and metabolites and enzymes related to oxidative stress were measured. In the asthmatics the concentration of glutathione, an antioxidant that protects cells from free radicals, was reduced and the concentration of the oxidized form (glutathione disulfide) was increased. These are the findings that would be predicted to occur during a period of oxidative stress.

Air Pollution and Chronic Obstructive Pulmonary Disease
Smoking tobacco is the most important risk factor for the development of COPD. First-hand and second-hand cigarette smoke are both risk factors. Most authors report that approximately 75% to 85% of all cases (as opposed to deaths, discussed above) of COPD can be attributed to this single, preventable cause. Data that have emerged during the past several years have shown that there is a smaller but important link between air pollution, including pollutants produced by burning coal, and the subsequent development of COPD.

In a study of the residents of Helsinki, Finland, pooled asthma and COPD emergency room visits increased on days when there were increases in the concentration of particles smaller than 2.5 μm in diameter ($PM_{2.5}$), larger particles and gaseous pollutants [24]. The Atlanta, Georgia, study of Peel et al. found that when the concentration of NO_2 or carbon monoxide increased by a standard deviation, emergency room visits for COPD increased by 2% to 3% [15]. Finally, in a study of hospitalization rates among Medicare enrollees, a 10 μg/m³ increase in the concentration of $PM_{2.5}$ particles was associated with a same-day increase in COPD admissions of 2.5% (95% CI 22.1–3.2%) [25]. These three studies of three different populations using different criteria, all link increases in air pollutants to exacerbations of COPD. Although they did not focus on pollutants derived exclusively from the combustion of coal, the pollutants they studied included those produced by coal burned by electrical utilities as well other sources.

Cancer of the Lung and Bronchus

Epidemiology and Demographics of Lung Cancer
The National Cancer Institute estimates that 222,520 Americans will develop lung cancer and that 157,300 will die from this disease in 2010 [26]. Data from 2007 indicate that cancers of the lung and bronchus are

ranked as the third most common site, trailing prostate, the most common, and female breast cancers [27]. For that year the overall cancer rate for Americans was 65.6 per 100,000: in men the incidence was 80.5 per 100,000, and for women it was 54.5 per 100,000. Rates by state varied substantially. Kentucky had the highest rate of 123.9 while Utah had the lowest at 32.0 [27]. For additional details, see figure 8.1. Rates were similar in whites and blacks (66.1 and 68.2 per 100,000, respectively) and lowest in Hispanics at 33.4 per 100,000.

Basic Lung Cancer Facts
The word cancer is derived from the Greek word for crab, which is *karkinos*. This is derived from the multi-armed growth of many cancers that may be seen as the cancer invades adjacent normal tissues. Cancers are due to the uncontrolled rapid growth and spread of cells. When viewed under a microscope, the size and shape of cancer cells is highly abnormal. Often the features of the tissue of origin are lost. Cancer cells are often bizarre in their size and shape. Mitotic figures, evidence for cell division, may be seen, particularly in the most malignant tumors. Cancers spread by direct extension into adjacent tissues or by metastasizing to distant organs. Most lung cancers arise from the epithelial cells. Lung cancer commonly spreads or metastasizes to the liver, bones, brain, and other organs.

The symptoms of lung cancer are shortness of breath, cough, particularly coughing up blood, and weight loss. Some lung cancers are found on routine chest X rays or by computed X-ray tomography, which has gained prominence as a screening examination. Lung cancers are usually either small-cell carcinomas or non–small-cell carcinomas. The distinction is important because treatments are very different.

Smoking tobacco and inhaling second-hand tobacco smoke are by far, the most important factors for the development of lung cancer. However, exposure to asbestos, arsenic, radon, and other radioactive gases; nickel compounds, and other airborne organic compounds; and air pollution have also been identified as risk factors.

Air Pollution and Lung Cancer
Data from three large epidemiological studies show clearly that air pollution must be considered as a risk factor for the development of lung cancer.

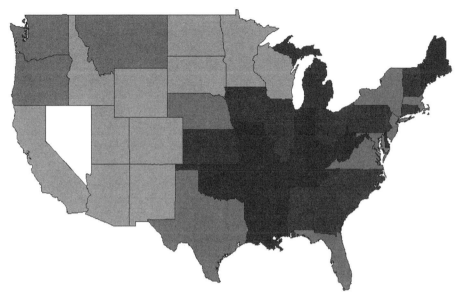

Color	Interval	States
	26–59.5	Arizona, California, Colorado, District of Columbia, Hawaii, Idaho, Minnesota, New Mexico, North Dakota, South Dakota, Utah, Wisconsin, and Wyoming
	59.6–66.3	Alaska, Florida, Maryland, Massachusetts, Montana, Nebraska, New Jersey, New York, Oregon, Texas, Virginia, and Washington
	66.4–74.7	Alabama, Connecticut, Georgia, Illinois, Iowa, Kansas, Michigan, New Hampshire, North Carolina, Ohio, Pennsylvania, South Dakota, and Vermont
	74.8–97.7	Arkansas, Delaware, Indiana, Kentucky, Louisiana, Maine, Mississippi, Missouri, Oklahoma, Rhode Island, Tennessee, and West Virginia
	Did not meet data quality criteria	Nevada

Figure 8.1
Lung and bronchial cancer prevalence by state. The prevalence of carcinomas of the lung and bronchus are indicated by shading on the grayscale. Source: Centers for Disease Control [37].

First among these was a study of Seventh Day Adventists who lived in California [28]. This group was selected because their religion discourages smoking and drinking alcohol. The investigators enrolled over 6,300 nonsmoking white adults and monitored them for the development of lung cancer from 1977 to 1992. These data were combined with monthly ambient air pollution data from appropriate zip codes. Positive associations with lung cancer were found for exposure to 10 μm particles, sulfur dioxide, and ozone. Since the study was published in 1998 and therefore preceded the 1997 standards for 2.5 micron particles, only the 10 μm particle data were available for analysis. The interquartile range, which is the difference between the first quartile and the third for the ozone concentration, was 100 ppm. For men, this was associated with an increase in the relative risk for lung cancer of 3.56 (95% CI, 1.35–9.42). This difference in ozone concentration was not associated with a significant effect in women. The effect of 10 μm particulate exposure was important among men and women. For men, the interquartile difference of 50 μg/m^3 was associated with a relative risk increase of 5.21 (95% CI = 1.94–13.00), whereas for women, the relative risk for this difference trended toward an increase of 1.21 (95% CI = 0.55–2.66). For sulfur dioxide, the relative risk associated with an interquartile concentration increase was 2.66 (95% CI = 1.62–4.39) for men and a relative risk increase of 2.14 (95% CI = 1.36–3.37) for women. Although these results were corrected for self-reports of smoking, drinking alcohol, and educational level, it is likely that the first two of these variables were underreported in these participants because their religious beliefs prohibit smoking and alcohol consumption. This could change the results, but the authors thought that this was not an important problem. The differences between men and women were thought to be due to greater exposures among men associated with the fact that men spend more time outside than women and were likely to have exercised more while outside, thereby increasing the amount of air they inhaled and thereby their exposure to air pollutants.

In the Harvard Six Cities study, the adjusted all-cause morality rate ratio for Steubenville, Ohio, the most polluted city, to Portage, Wisconsin, the least polluted city was 1.26 (95% CI, 1.08–1.47). Lung cancer accounted for 8.4% of all deaths. After adjustment for cigarette smoking, body mass index, educational level, and other health-related factors, the cancer risk in Steubenville was 1.37 times higher than the risk in Portage (95% CI = 0.81–2.31) [29].

Complementary data were found in the American Cancer Society study [30]. This epidemiological study began with 1.3 million adults in 1982. From that set, approximately 500,000 adults were matched with air pollution data for their appropriate metropolitan area and vital statistics data through the end of 1998. Fine particulate increases of 10 μg/m³ were associated with an 8% increase in lung cancer mortality. Increases in the concentration of oxides of sulfur were also associated with increases in lung cancer mortality.

These three large prospective epidemiological studies provide convincing evidence that air pollution, particularly that due to particulates and ozone, has substantial effects on mortality due to lung cancer.

Coal Workers' Pneumoconiosis or Black Lung Disease

Epidemiology and Demographics of *Coal Workers' Pneumoconiosis*

Black lung disease is the common name for coal workers' pneumoconiosis. It is a chronic disease of the lungs that is due to the inhalation of dust from coal mines. Both surface and underground miners are at risk. The Coal Mine Health and Safety Act of 1969 established limits on the exposure to dust from coal mines and, simultaneously, established a chest X-ray screening program for underground miners. As a result of this program, the prevalence of this disease fell between the 1970s and the 1990s. However, since then data from chest X-rays taken as a result of the Act's surveillance program show that the prevalence has been increasing [31]. Seaton attributed this increase to a relaxation of regulatory activity due to lobbying efforts of "the wealthy who control the media" [32].

As one might expect, the prevalence of coal workers' pneumoconiosis increases as the number of years spent mining increases [33]. Among just over 31,000 miners who were surveyed with chest X-rays, the overall prevalence was 2.8%, rising from 0.2% among miners who were less than 30 years old to 5.1% among miners who were 60 or more years of age. Among those with 9 or fewer years of underground mining experience the prevalence was 1.1%. This rose to 5.4% for those with 25 or more years of experience. The prevalence was lower among surface miners. Among those with 9 or fewer years of experience the prevalence was 0.6%, rising to 3.4% among those with 25 or more years of experience. Progressive massive fibrosis (see below) is a more severe form of the disease and affects 0.3% or fewer of coal miners. Again, length of exposure and underground versus aboveground mining are modifiers of the risk for this disorder. Mine size is also a factor [31]. Both forms of the disease are significantly more prevalent among miners who work underground in mines that employ fewer than 50 workers compared to their colleagues who work in larger mines.

Basic Facts

Coal workers' pneumoconiosis is a chronic disease of the lungs that is present in coal miners. The diagnosis is based on three criteria: a chest X ray that is consistent with the condition, a work history that indicates the exposure to coal dust is of sufficient duration, and elimination of other illnesses that could mimic the disease [34]. Pathologically, coal workers' pneumoconiosis is characterized by focal collections of coal dust in pigment-laden macrophages, sometimes known as the body's garbage collectors. These collections are known as coal macules and are the characteristic lesion of the disease. These macules are usually associated with an enlargement of adjacent air spaces known as centrilobular emphysema (i.e., it occurs in bronchioles, very small terminal portions of the airway). As macules expand, they may become nodules that may be felt when the lung is palpated. The diagnostic criteria for progressive massive fibrosis are met when one or more nodules that are 2 centimeters or greater in size are present [34]. Aside from modifying risk factors, such as smoking cigarettes and mining, there is no specific treatment.

Cough and shortness of breath are the primary symptoms. When the disease progresses, there may be secondary effects on the heart known as cor pulmonale. Cor pulmonale may also be present in patients with other forms of chronic lung disease, such as COPD, and commonly leads to heart failure and death. Occasionally the nodules will rupture and spill their contents into the airway causing black tinting of sputum.

At one time, it was thought that the silicon in coal dust was the factor responsible for the development of this disease. However, more recent evidence suggests that the amount of iron in the coal dust may be of critical importance [35,36]. Iron may facilitate the formation of free radicals that are elements of the inflammatory and oxidative stress pathogenetic factors.

Summary

It should not be too surprising that coal-derived air pollution has many direct effects on the lungs. This is shown by multiple, well-designed, epidemiological studies showing correlations between the pollutants and diseases of the lower airways, including asthma, COPD, and lung cancer. The cause-and-effect relationship is strengthened by the fact that these studies have been replicated in numerous settings, across various age groups, and in multiple countries. Further strength comes from the fact that there are plausible mechanisms that can explain the relationships and from various studies in laboratories where controlled exposures to human volunteers and animals can be conducted.

References

1. Miniño AM, Xu J, Kochanek KD. Deaths: Preliminary Data for 2008. Atlanta: CDC, 2010.

2. Moorman JE, Zahran H, Truman BI, Molla MT. Current asthma prevalence— United States, 2006–2008. MMWR—Morbidity and Mortality Weekly Report 58(50):1412–6, 2011;60(suppl):84–6.

3. Akinbami LJ, Moorman JE, Garbe PL, Sondik EJ. Status of childhood asthma in the United States, 1980–2007. Pediatrics 2009;123 (suppl 3):S131–45.

4. Centers for Disease Control. Behavioral Risk Factor Surveillance System Prevalence Data for Asthma. Available at http://www.cdc.gov/asthma/brfss/default.htm.

5. Yang IA, Fong KM, Zimmerman PV, Holgate ST, Holloway JW. Genetic susceptibility to the respiratory effects of air pollution. Thorax 2008;63(6):555–63.

6. Bateson TF, Schwartz J, Bateson TF, Schwartz J. Children's response to air pollutants. J Toxicol Environ Health Part A 2008;71(3):238–43.

7. Trasande L, Thurston GD, Trasande L, Thurston GD. The role of air pollution in asthma and other pediatric morbidities. J Allergy Clin Immunol 2005;115(4):689–99.

8. Adams WC. Comparison of chamber and face-mask 6.6-hour exposures to ozone on pulmonary function and symptoms responses. Inhal Toxicol 2002;14(7):745–64.

9. Adams WC. Comparison of chamber 6.6-h exposures to 0.04–0.08 PPM ozone via square-wave and triangular profiles on pulmonary responses. Inhal Toxicol 2006;18(2):127–36.

10. Brown JS, Bateson TF, McDonnell WF. Effects of exposure to 0.06 ppm ozone on FEV_1 in humans: a secondary analysis of existing data. Environ Health Perspect 2008;116(8):1023–6.

11. Schelegle ES, Morales CA, Walby WF, Marion S, Allen RP. 6.6-hour inhalation of ozone concentrations from 60 to 87 parts per billion in healthy humans. Am J Respir Crit Care Med 2009;180(3):265–72.

12. US Environmental Protection Agency. Provisional assessment of recent studies on health and ecological effects of ozone exposure. Report EPA/600/R-09/101. Washington DC: EPA, 2009.

13. US Environmental Protection Agency. Air quality criteria for ozone and related photochemical oxidants. Report EPA/600/R-05/004aF. Washington DC: EPA, 2006.

14. Gent JF, Triche EW, Holford TR, et al. Association of low-level ozone and fine particles with respiratory symptoms in children with asthma. JAMA 2003;290(14):1859–67.

15. Peel JL, Tolbert PE, Klein M, et al. Ambient air pollution and respiratory emergency department visits. Epidemiology 2005;16(2):164–74.

16. Gilmour MI, Jaakkola MS, London SJ, et al. How exposure to environmental tobacco smoke, outdoor air pollutants, and increased pollen burdens influences the incidence of asthma. Environ Health Perspect 2006;114(4):627–33.

17. Brauer M, Hoek G, van VP, et al. Air pollution from traffic and the development of respiratory infections and asthmatic and allergic symptoms in children. Am J Respir Crit Care Med 2002;166(8):1092–8.

18. Gehring U, Cyrys J, Sedlmeir G, et al. Traffic-related air pollution and respiratory health during the first 2 yrs of life. Eur Respir J 2002;19(4):690–8.

19. McConnell R, Berhane K, Gilliland F, et al. Asthma in exercising children exposed to ozone: a cohort study. Lancet 2002;359(9304):386–91.

20. Ritz B, Wilhelm M, Zhao Y. Air pollution and infant death in southern California, 1989–2000. Pediatrics 2006;118(2):493–502.

21. Roberts ES, Richards JH, Jaskot R, Dreher KL. Oxidative stress mediates air pollution particle-induced acute lung injury and molecular pathology. Inhal Toxicol 2003;15(13):1327–46.

22. Rhoden CR, Ghelfi E, Gonzalez-Flecha B. Pulmonary inflammation by ambient air particles is mediated by superoxide anion. Inhal Toxicol 2008;20(1):11–5.

23. Fitzpatrick AM, Teague WG, Holguin F, Yeh M, Brown LA. Airway glutathione homeostasis is altered in children with severe asthma: evidence for oxidant stress. J Allergy Clin Immunol 2009;123(1):146–52.

24. Halonen JI, Lanki T, Yli-Tuomi T, Kulmala M, Tiittanen P, Pekkanen J. Urban air pollution, and asthma and COPD hospital emergency room visits. Thorax 2008;63(7):635–41.

25. Dominici F, Peng RD, Bell ML, et al. Fine particulate air pollution and hospital admission for cardiovascular and respiratory diseases. JAMA 2006;295(10): 1127–34.

26. National Cancer Institute. Lung Cancer. Atlanta GA, 2011.

27. National Program of Cancer Registries. United States Cancer Statistics: 2007 Top Ten Cancers. Atlanta: CDC, 2011.

28. Beeson WL, Abbey DE, Knutsen SF. Long-term concentrations of ambient air pollutants and incident lung cancer in California adults: results from the AHSMOG study. Adventist Health Study on Smog. Environ Health Perspect 1998;106(12):813–23.

29. Dockery DW, Pope CA, III, Xu X, et al. An association between air pollution and mortality in six U.S. cities. N Engl J Med 1993;329(24):1753–9.

30. Pope CA, III, Burnett RT, Thun MJ, et al. Lung cancer, cardiopulmonary mortality, and long-term exposure to fine particulate air pollution. JAMA 2002; 287(9):1132–41.

31. Laney AS, Attfield MD. Coal workers' pneumoconiosis and progressive massive fibrosis are increasingly more prevalent among workers in small underground coal mines in the United States. Occup Environ Med 2010;67(6):428–31.

32. Seaton A. Coal workers' pneumoconiosis in small underground coal mines in the United States. Occup Environ Med 2010;67(6):364.

33. Centers for Disease Control. Pneumoconiosis prevalence among working coal miners examined in federal chest radiograph surveillance programs—United States, 1996–2002. MMWR—Morbidity and Mortality Weekly Report 58(50): 1412–6, 2003;52(15):336–40.

34. Banks DE. Coal workers' pneumoconiosis. In: Schwartz MI, King TE Jr., eds, Interstitial Lung Disease, 4 ed. Hamilton NY: Decker, 2003;402–17.

35. Huang X, Li W, Attfield MD, Nadas A, Frenkel K, Finkelman RB. Mapping and prediction of coal workers' pneumoconiosis with bioavailable iron content in the bituminous coals. Environ Health Perspect 2005;113(8):964–8.

36. McCunney RJ, Morfeld P, Payne S. What component of coal causes coal workers' pneumoconiosis? J Occup Environ Med 51(4):462–71, 2009.

37. Centers for Disease Control. Lung Cancer Rates by State. Atlanta: CDC, 2007.

9
Diseases of the Cardiovascular System

Diseases of the heart were the leading cause of death in the United States in 2008, according to the preliminary analysis compiled by the Centers for Disease Control [1]. Even though this represented a 2.2% decrease from 2007, 617,527 Americans died from heart disease in that year. This represents a rate of 203.1 deaths per 100,000 individuals. For those over 65, the rate was substantially higher, at 1,277.8 per 100,000.

The American Heart Association estimates that 82.6 million Americans, or about one out of every three Americans, has some form of cardiovascular disease, according to 2007 data [2]. This is a relatively broad category of diseases and includes hypertension, coronary heart disease, heart failure, and others. The number of Americans affected by these diseases is shown in table 9.1. The prevalence of heart disease is slightly higher among whites than blacks, 11.9% and 11.2%, respectively. Hypertension is much more prevalent among blacks than among whites, 32.2% compared to 23%, respectively. Cardiovascular disease deaths were the lowest in Minnesota where the rate was 193.1 per 100,000 and highest in Mississippi, where the rate was 349.7 per 100,000. The total direct costs of heart disease were thought to be $82.2 billion, with another $95.3 in indirect costs due to premature mortality and lost productivity. Any way you look at heart disease, it is a huge source of death and disability.

Atherosclerotic Cardiovascular Disease

The control of risk factors has made the most important contribution to the decline in death rates attributable to heart disease over the past

Table 9.1
Cardiovascular disease prevalence in the US population, 2007 data [2]

Disease	Prevalence
Hypertension	76,400,000
Coronary heart disease	16,300,000
Acute myocardial infarct	7,900,000
Angina pectoris	9,000,000
Congestive heart failure	5,700,00

decades. The American Heart Association and other organizations have issued guidelines designed to aid health professionals as they seek to keep heart disease from developing (primary prevention) and to prevent progression of disease to avoid later more severe manifestations (secondary prevention). Traditionally these guidelines have focused on the control of hypertension, cholesterol levels, smoking, and other factors. More recently they have been expanded to deal with life-style choices, such as diet, exercise, and avoidance of second-hand smoke. However, because of accumulating evidence and a persistent concern that air pollutants are also linked to adverse cardiovascular health outcomes, the AHA convened an expert panel to evaluate this threat. The results of their deliberations, the single most authoritive review of this topic, were published in 2004 and expanded and revised in 2010 [3,4]. Chapter 6 and figure 6.2 summarize the pathophysiological mechanisms by which air pollutants, especially particulate matter, cause cardiovascular disease. The figure illustrates how inhaled particulate matter affects blood and the autonomic nervous system and induces systemic oxidative stress and inflammation. These effects, in turn, trigger additional, more specific effects that lead to the development of diseases of the heart and circulatory system. The details of these processes are extremely complex and beyond the scope of this book (see chapter 6 and AHA reports for additional details [3,4]). The particulates with the smallest aerodynamic diameter are the greatest concern. By convention, and for purposes of monitoring air to evaluate compliance with air quality standards, the particulates with the greatest effects on health have a diameter of 2.5 microns (μm) or less. These small particles are the most likely to penetrate deeply into the lungs, reach the alveoli, and initiate the pathophysiological sequences leading to acute and chronic manifestations of heart disease.

These mechanisms suggest numerous possible therapeutic interventions; reducing the exposure to airborne pollutants is the most obvious of these. This is in accord with one of the fundamental principles of medicine: to prevent is better than to cure.

Impact of Air Pollutants on the Cardiovascular System

The impacts of air pollutants on the cardiovascular system can be separated into two categories: immediate impacts and long-term impacts. While both are important, different methods must be employed, different conclusions may be reached, and differences in policies may result.

Immediate Impacts

Acute outcome studies typically focus on a single event, such as an admission to the hospital or emergency room due to acute myocardial infarction or the discharge of an implanted cardioverter-defibrillator. These events have a distinct time of occurrence that can be found in hospital emergency room records, examination of defibrillator data extracted after a discharge, or some similar source. These data are then evaluated in relationship to air quality data typically collected by monitoring stations. Geographical information, such as zip codes, is used to link the data sets during the statistical analyses that ensue. Access to large data sets, such as Medicare records, combined with the data from nationwide air quality monitoring stations, developed as required by the Environmental Protection Agency (EPA) or equivalent state agencies, have enabled epidemiologists to acquire the large numbers of cases required to identify statistically significant associations between air quality and an event. These associations are highly relevant to the EPA mission: to protect health and the environment.

Because of studies that linked air pollution to hospital admissions for cardiovascular disease, Peters et al. hypothesized that there might be a link between transient increases in pollutant levels and therapeutic discharges of implanted defibrillators [5]. These devices are implanted permanently in patients judged to be at risk for sudden death due to cardiac arrhythmias. These devices monitor heart rhythms continuously. When a serious, potentially fatal rhythm disturbance, such as ventricular fibril-

lation or ventricular tachycardia, is detected, the device begins to pace
and/or defibrillate the heart. The goal is to restore a normal heart rhythm.
Defibrillation is transient but quite painful. After a defibrillator discharge
patients are instructed to seek medical attention. Modern devices typically include memory chips that store information, including the time of
the event and the associated electrocardiogram. When the patient seeks
medical attention after a discharge, technicians are able to retrieve
relevant data, including the nature and time of the arrhythmia and
the discharge. Peters and her colleagues analyzed the records from 100
events recorded in a single clinic in eastern Massachusetts and sought
links between events and peaks in pollutant levels measured in that
region. They considered daily average pollutant levels on the day of
the event and levels 1, 2, and 3 days prior to the event. They found that
an increase in the nitrogen dioxide concentration was followed by an
increase in the probability of a discharge 2 days later (odds ratio 1.8,
95% CI = 1.1–2.9). Patients who experienced 10 or more discharges
(presumably an indication of more severe disease) exhibited associations
with nitrogen dioxide, carbon monoxide, black carbon, and particulates
with a diameter of 2.5 μm or less. Although the authors regarded this
as a pilot study, they concluded that peaks in air pollution levels
were associated with fatal or potentially fatal cardiac arrhythmias. To
buttress their claim, they reviewed the results of animal studies linking
pollutants to cardiac arrhythmias.

Peters and her colleagues also investigated the relationship between
acute heart attacks or myocardial infarctions (MI) and air pollutants [6].
They reviewed a total of 772 records of patient interviews conducted
within 4 days of an acute MI. Examining actual patient histories minimizes bias that could be introduced by asking patients to remember a
distant event (a phenomenon known as recall bias). They also retrieved
air pollution data associated with the date of the event. They paired
the data from the day of the MI with data from three other days when
the subject did not have an MI. This way the patients served as their
own controls. Compared to control periods, there was an increase in the
probability of an MI in association with small-diameter particle levels
measured two hours before the MI (the odds ratio for an increase of
25 μg/m^3 was 1.48 with 95% CI = 1.09–2.02). In addition there was a
delayed response to a peak occurring a full day before an event (the

odds ratio for the same increase in the concentration was 1.69, with 95% CI = 1.13–2.34).

Two large studies, using health outcomes such as mortality in relation to day-to-day changes in ambient air pollution levels, have been critical in defining the health effects associated with pollutants. In a US study Dominici et al. used hospital admission rates in the Medicare National Claims History Files and ambient small particle concentrations to look for associations between particulate levels and admissions for ischemic heart diseases, disturbances of heart rhythm, and congestive heart failure [7]. The data were extracted from 204 urban counties with a total of 11.5 million Medicare enrollees who lived an average of 5.9 miles from a monitor that measured the concentration of 2.5 µm particles. Compared to injuries, for which there was no reason to expect an effect of air pollution (i.e., they used injuries as a control), they found increases in all categories, with the largest found for congestive heart failure where a 1.28% increase in admissions occurred with an increase of 10 µg/m^3 in the small particle concentration. Increases in admissions for ischemic heart disease, heart failure, and disturbances of heart rhythm tended to be higher among those 75 years old or greater than among those aged 65 to 74. Additional details are shown in table 9.2. The greatest effects were observed in the northeastern United States where coal-fired power plants are most plentiful. Although the increases in the rates appear small, on the order of a percent, the large number of Medicare enrollees translates the result into a very large effect, when measured in terms of total hospital admissions, patient morbidity and mortality, and the cost of health care and lost opportunities.

Table 9.2
Percent change in hospitalization rate per 10 µg/m^3 increase in PM$_{2.5}$ for all Medicare enrollees age > 65 years [14]

Admission diagnosis	Lag days[a]	Percent rate increase (95% confidence interval)
Ischemic heart disease	2	0.44 (0.02–0.86)
Heart rhythm abnormality	0	0.57 (-0.01–1.15)
Heart failure	0	1.28 (0.78–1.78)

a. Number of days between peak and greatest effect of PM$_{2.5}$, particles with an aerodynamic diameter of 2.5 µm or less.

Katsouyanni et al. reported the short-term effects of 10 μm particles on the health of the residents of 29 European cities [8]. The study extended over a period of about five years and included approximately 43 million people. Unlike the US study, these investigators included all age groups. However, the results were remarkably similar to those observed in the US Medicare population. They reported a 0.6% increase (95% CI = 0.4–0.8%) in the daily number of deaths for a 10 μg/m^3 increase in the concentration of the particles. They also found important modifiers of the death rate attributed to particles, particularly with regard to nitrogen dioxide levels. Cities with high concentrations of this gas had death rates that were approximately four times higher than found in cities with low concentrations. Death rates in cities with warm climates were about 2.8 times higher than in cities with cold climates.

These examples that describe the impact of air pollution on acute morbidity and mortality are remarkably consistent. The small studies focused on individuals as well as large studies that rely on data extracted from huge databases show important adverse effects of pollutants on indicators of acute cardiovascular illness.

Long-Term Impacts

Two studies linking the chronic effects of air pollutants on cardiovascular mortality are particularly relevant. The first of these is the Harvard Six Cities study, reported by Dockery et al. [9]. This was a prospective cohort study in which the investigators followed over 8,100 adults for 14 to 16 years. The mortality rate in the most polluted city was 1.26 times higher than the rate in the least polluted city (95% CI = 1.08–1.47). This elevated rate persisted after controlling for important life-style confounders, including smoking cigarettes.

In a another quite recent study published in *The New England Journal of Medicine*, perhaps the leading medical journal in the world, the authors reported on the effect of changing particulate concentrations on overall mortality [10]. In this retrospective study, the investigators retrieved particulate air pollution data from 1979 to1983 and in the 1997 to 2001 time intervals. They linked these data with overall mortality as revealed by vital statistics and cause of death data. The study focused on 211 counties encompassed by 51 metropolitan areas. In the study interval, particulate concentrations decreased. They used multiple models to evaluate the data, controlling for a variety of risk factors, including smoking cigarettes and socioeconomic status. A decrease in the fine particulate matter concentration of 10 µg/m^3 was associated with a 6% decrease in the risk of death due to cardiopulmonary causes. The improvements in air quality accounted for about 15% of the improved overall mortality in these metropolitan areas. Some cities, such as Buffalo, New York, were big winners. In Buffalo, the small particle concentration fell by 13 µg/m^3 and life expectancy rose by 3.4 years. It was during this time that the steel industry collapsed in Buffalo and the use of coal fell dramatically.

In the report of the Women's Health Initiative, a prospective cohort study of over 65,000 postmenopausal women, there were 1816 cardiovascular events in the follow-up period [11]. Their data showed that for each 10 µg/m^3 increase in the concentration of particulate matter with an aerodynamic diameter of 2.5 µm or less there was a 24% increase in the risk of a cardiovascular event (hazard ratio = 1.24, 95% CI = 1.09–1.41) and a 76% increase in the risk for death due to a cardiovascular event (hazard ratio = 1.76, 95% CI = 1.25–2.47).

These studies show that air pollutants from coal combustion have serious long-term impacts on the cardiovascular health of the US population.

Controlled Studies of Humans and Animals

The strongest, most reliable medical evidence comes when pooling data from multiple randomized controlled trials—a meta-analysis. Single randomized controlled trials are the next strongest form of evidence. As one might imagine randomized controlled trials involving air pollutants are difficult to perform and pose clear ethical dilemmas. Is it reasonable to deliberately expose individuals to air pollutants that are known to have adverse effects on health?

Despite the barriers to performing gold standard trials, there are some that support the hypothetical pathogenetic mechanisms outlined in chapter 6. Typically these exposure studies use normal healthy volunteers. The risk of a serious effect in this population is typically similar to the risks experienced in many cities on days when air pollution is at its worst.

One of these randomized controlled trials was conducted by Brook et al. who evaluated the effects of fine particles and ozone on the diameter of an artery in the arm (the brachial artery) in 25 healthy adults [12]. These arteries are close to the surface of the skin and their behavior is thought to be representative of the behavior of coronary and cerebral arteries. In the study the investigators evaluated the cardiovascular response to a two-hour inhalation of fine particles (approximately 150 micrograms per cubic meter) and ozone (120 parts per billion). These concentrations are similar to those that may be encountered in many urban settings. Initially the participants were randomly assigned to inhale air or the pollutant mixture. Neither the participants nor the investigators were aware of this initial assignment. After this first experimental period, the participants were switched from the pollutant mix to normal air, and vice versa. In technical terms, this was a double-blind, randomized, crossover study, an ideal design for a clinical trial. The investigators used ultrasound devices to evaluate the diameter of the artery during the exposure, a harmless, noninvasive technique. They found that the exposure to the two pollutants resulted in a significant reduction in the diameter of the brachial artery, one of the major arteries in the arm. This implied that there was likely to be a similar narrowing of other arteries, such as those in the heart or brain. Despite this clear-cut evidence, important questions remain; for example, are these participants representative of the population at greatest risk (screened healthy controls versus patients with significant coronary artery disease)? Because of these limitations, epidemiological data and observational methods must be used to study populations. These observational studies deal with the world as it exists, including vulnerable segments of the population, rather than the more artificial laboratory environment where ethical constraints mandate the selection of research participants who are not vulnerable.

Animal studies provide greater flexibility in terms of the experimental conditions employed and measurement of endpoints. They also provide support for the observations made in epidemiological studies. As described in chapter 6, pulmonary inflammation and the presence of reactive oxygen species (ions, free radicals formed from oxygen) are thought to be important in the pathogenesis of cardiac disease. Animal studies are well suited to studying these mechanisms.

In a study of rabbits specially bred to have high concentrations of lipids in their blood, Suwa et al. found that a four-week exposure to PM_{10} (particulate matter with an aerodynamic diameter of 10 μm or less) was associated with increase in the rate of development of atherosclerosis, an increase in the turnover of cells in atherosclerotic plaques, and an increase in the total lipid content of aortic lesions [13].

Summary

The short- and long-term exposure to air pollutants, particularly to fine particles, is a major risk factor for the development of cardiovascular diseases (CVD). The strength of the epidemiological evidence linking particulates with an aerodynamic diameter of 2.5 μm or less to CVD was summarized by an American Heart Association consensus committee [4]. For exposures on the order of days, there is a very strong link to cardiovascular deaths, and hospitalizations, and fatal and nonfatal episodes of ischemic heart disease (heart attacks and angina pectoris). The link to fatal and nonfatal exacerbations of heart failure and ischemic strokes was not quite as strong. Limited evidence linked this short exposure to episodes of other vascular diseases and heart rhythm abnormalities, including cardiac arrest. Longer term exposures, on the order of months to years, is linked strongly to cardiovascular deaths and hospitalizations as well as fatal and nonfatal episodes of ischemic heart disease. More limited evidence linked these longer exposures to heart failure, ischemic stroke, vascular diseases, and cardiac arrhythmias including cardiac arrest. The strength of the evidence for excess mortality is the result of several large studies, reviewed above, that included diverse populations. Diseases of the cardiovascular system lead all others as a cause of death among Americans and are a leading cause of death worldwide. The studies reviewed in this chapter show clearly that pollutants produced by burning coal have powerful effects on physiological processes that lead to cardiovascular diseases and death.

References

1. Miniño AM, Xu J, Kochanek KD. Deaths: Preliminary Data for 2008. Atlanta: Centers for Disease Control, 2010.

2. Roger VL, Go AS, Lloyd-Jones DM, et al. Heart disease and stroke statistics—2011 update: a report from the American Heart Association. Circulation 2010.1;123(4):e18–209.

3. Brook RD, Franklin B, Cascio W, et al. Air pollution and cardiovascular disease: a statement for healthcare professionals from the Expert Panel on Population and Prevention Science of the American Heart Association. Circulation 2004;109(21):2655–71.

4. Brook RD, Rajagopalan S, Pope CA III, et al. Particulate matter air pollution and cardiovascular disease: an update to the scientific statement from the American Heart Association. Circulation 2010;121(21):2331–78.

5. Peters A, Liu E, Verrier RL, et al. Air pollution and incidence of cardiac arrhythmia. Epidemiology 2000;11(1):11–7.

6. Peters A, Dockery DW, Muller JE, Mittleman MA. Increased particulate air pollution and the triggering of myocardial infarction. Circulation 2001;103(23): 2810–15.

7. Dominici F, Peng RD, Bell ML, et al. Fine particulate air pollution and hospital admission for cardiovascular and respiratory diseases. JAMA 2006;295(10): 1127–34.

8. Katsouyanni K, Touloumi G, Samoli E, et al. Confounding and effect modification in the short-term effects of ambient particles on total mortality: results from 29 European cities within the APHEA2 project. Epidemiology 2001;12(5): 521–31.

9. Dockery DW, Pope CA III, Xu X, et al. An association between air pollution and mortality in six U.S. cities. N Engl J Med 1993;329(24):1753–9.

10. Pope CA III, Ezzati M, Dockery DW. Fine-particulate air pollution and life expectancy in the United States. N Engl J Med 2009;360(4):376–86.

11. Miller KA, Siscovick DS, Sheppard L, et al. Long-term exposure to air pollution and incidence of cardiovascular events in women. N Engl J Med 2007; 356(5):447–58.

12. Brook RD, Brook JR, Urch B, Vincent R, Rajagopalan S, Silverman F. Inhalation of fine particulate air pollution and ozone causes acute arterial vasoconstriction in healthy adults. Circulation 2002;105(13):1534–6.

13. Suwa T, Hogg JC, Quinlan KB, Ohgami A, Vincent R, van Eeden SF. Particulate air pollution induces progression of atherosclerosis. J Am Coll Cardiol 2002;39(6):935–42.

14. Dominici F, Peng RD, Bell ML, et al. Fine particulate air pollution and hospital admission for cardiovascular and respiratory diseases. JAMA 2006;295(10): 1127–34.

10

Diseases of the Nervous System

It is easy to understand that burning coal is likely to have an adverse impact on respiratory health—we inhale the products of combustion. It is less obvious that burning coal has important effects on the nervous system, particularly the brain. Cerebral vascular disease, namely stroke, and loss of intellectual capacity due to mercury, are the two most important neurological consequences of burning coal.

The Brain and Pollution

Our brain is the organ that most clearly distinguishes us from other species. Our abilities to use language, think abstractly, produce and enjoy music, art and literature, to inquire about the nature of the universe and a host of activities, all depend on our brains.

Our brain is constantly hard at work. At rest, between 15% and 20% of the blood pumped by the heart goes to the brain. The normal brain weighs 1,300 to 1,400 grams, or about 3 pounds, and has one of the highest metabolic rates of any organ in the body. Just as our automobile engines require a constant supply of gasoline and oxygen, the brain requires glucose and oxygen. Unlike a car, with its tank full of gasoline, the brain does not have a reserve supply of energy in the form of glucose. The combination of the extremely high metabolic rate of the brain and the absence of a way that the brain can store the energy sources it needs makes it necessary to deliver oxygen and glucose all of the time and at a high rate. Thus interruptions of the supply of blood, oxygen, or glucose may result in severe, permanent brain injury or worse, brain death. This is what happens when your blood pressure falls and you faint, or when an artery to the brain is blocked, and you have a stroke.

The complexity of the brain, coupled with its susceptibility to the effects of metabolic or physiological derangements or both, frequently leads to abnormalities of brain function. These may be very subtle and undetectable in an individual. However, if a small, seemingly insignificant deficit appears in a large number of individuals, epidemiological research methods may reveal an enormous adverse impact on a population as a whole. This, in turn, has public health consequences and an effect on the nation's future and its economy. This is illustrated in figure 10.1, where the impact of a 5 point decrement in IQ is depicted. The average IQ score is 100, and 95% of all individuals have IQ scores that fall between 70, the score below which one is considered to have an intellectual disability, and 130, the score above which one is considered to have superior intelligence. This is shown in panel a of figure 10.1. Panel b of figure 10.1 demonstrates the effect of an across-the-board 5 point decrement in IQ. For an individual, this relatively small change would almost certainly escape notice or it would be attributed to expected, day-to-day variations in performance. However, in a large population, substantial numbers of individuals are removed from the superior intelligence category and others are pushed down into the intellectually disabled category. The result is a smaller pool of individuals with outstanding intellects that move civilization ahead and a larger pool that requires many resources to be able to function. This is how intellectual capacity, or lack thereof, in a population has an effect on the economy, which, in turn, affects aspects of national security. It is this reality that makes it important to protect and preserve the brain's full potential.

Cerebral Vascular Disease

The National Institute of Neurological Diseases and Stroke says, "A stroke occurs when the blood supply to part of the brain is suddenly interrupted or when a blood vessel in the brain bursts" [1]. The term stroke does not define a single entity. It refers to a variety of acute cerebrovascular events including: ischemic stroke, caused by occlusion of a cerebral artery by an atherosclerotic plaque or an embolus; cerebral hemorrhage, usually caused by rupture of a small artery in the brain; and subarachnoid hemorrhage, typically due to rupture of an aneurysm or a malformation of blood vessels in the brain.

(a)

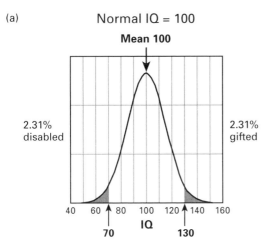

(b)

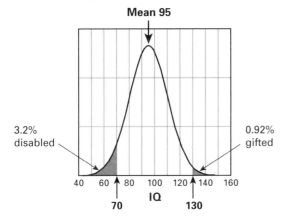

Figure 10.1
Effects of IQ loss on a population. (a) In a population with a mean IQ of 100, about 2.31% are intellectually disabled and an equal number are gifted. (b) A five-point IQ reduction shifts the curve to the left so that about 3.2% are now intellectually disabled and 0.92% are gifted.

The same pathophysiological mechanisms that affect the coronary arteries, shown in figure 6.2, causing myocardial infarcts, also apply to the arteries that nourish the brain. These common mechanisms include stimulation of the inflammatory response in cerebral vessels leading to atherosclerotic plaque formation, rupture, and arterial occlusion; oxidative stress; and alterations in blood viscosity. In addition cardiac arrhythmias may cause transient reductions of the cardiac output and a reduction of blood flow to the brain. The effects of reductions in the cardiac output are most prominent in the regions of the brain that are found at the boundaries between major arteries (so-called watershed areas of the brain), and at brain regions that are downstream from very narrow arteries.

Epidemiology and Demography of Stroke

Although there have been major improvements in primary and secondary prevention of strokes in the past several decades, stroke is still ranked fourth as a cause of death among Americans [2]. The stroke death rate in the United States fell 26.9% between the intervals of 1999 to 2001 and 2005 to 2007 [3]. The Centers for Disease Control reports over 133,000 stroke deaths among Americans in 2008, a decrease of 3.8% from the number reported in 2007. These reductions are due to many factors, including better control of blood pressure among patients with hypertension, widespread use of drugs to control blood cholesterol levels, smoking cessation efforts, and other preventive measures. The stroke death rate for men is 33.1 per 100,000 and 26.1 per 100,000 for women: the risk among blacks is about twice the risk among whites [4]. New Yorkers have the lowest stroke death rate at 29.6 per 100,000 and Arkansans have the highest at 58 per 100,000 [3].

Air Pollution and Stroke

Three large epidemiological investigations and several smaller studies have shown clear relationships between air pollutants and acute strokes.

In their study of fine particles, those with a diameter of 2.5 microns or less, and hospital admission rates in the Medicare population, Domenici et al. reported that for an increase in the concentration of these particles of 10 µg/m^3, there was a 0.81% increase in the hospitalization rate for most common diagnostic codes for cerebrovascular disease (95% CI = 0.31–1.32) [5]. (Note: Diagnostic codes, developed by the International Statistical Classification of Diseases and Related Health Problems, are used to develop health care statistics and determine reimbursement rates for various diseases.) In that study they also examined the delay between a peak in the concentration of the particles and the hospitalization rate. The effect of a peak was confined to the day that the peak occurred, that is, lag day 0 (no lag between a peak and the admission). There was no peak effect on lag days 1 and 2. This association was strongest in eastern parts of the United States when compared to western regions.

There may be several explanations for this regional effect. The mean concentration of fine particles is higher in this region than elsewhere, as shown in figure 3.3, and particulates in this region contain more sulfates [6]. Particulates are critical determinants of many health care outcomes, including mortality and hospital admissions, as shown in table 12.4. The contribution of particulate emissions to the hidden costs of burning coal to generate electricity is portrayed vividly in the report by the National Research Council described in chapter 12 and illustrated in figure 13.2.

In a second study of the Medicare population, Wellenius et al. looked for relationships between hospital admissions from 1986 to 1993 for ischemic and hemorrhagic strokes and increases in various air pollutants in nine major US cities [7]. Admissions data were obtained from the Centers for Medicare and Medicaid Services and pollutant levels were obtained from the EPA. Pollutant levels and admission data are shown in table 10.1. Note that the concentration of particles with a diameter of 10 μm or less was measured, and not the fine particle fraction as would be the case for more contemporary studies. They report that for an increase in the concentration from the 25th percentile to the 75th percentile in the PM_{10} concentration, there was a corresponding 1.03% increase in the admissions for ischemic stroke (95% CI = 0.04–2.04%) on the day of the increase. Similar results were observed for carbon monoxide and oxides of nitrogen and sulfur for ischemic strokes only. They did not find significant associations between pollutant levels and hemorrhagic strokes.

Finally, data from postmenopausal women enrolled in the Women's Health Initiative collected in 2000 show that for each increase of 10 μg/m³ in the $PM_{2.5}$ concentration (particles with a diameter of 2.5 μm or

Table 10.1
Effects of pollutant concentration on stroke admissions

Pollutant	Interquartile concentration increase	Percent increase in hospitalization rate with 95% CI
PM_{10}	22.96 µg/m^3	1.03 (0.04–2.04)
Carbon monoxide	0.71 ppm	2.83 (1.23–4.46)
NO_2	11.93 ppb	2.94 (1.78–4.12)
SO_2	6.69 ppb	1.35 (0.43–2.29)

Source: Adapted from table 3 of Wellinius et al. [7].
Note: The percent change in hospital admission rate for ischemic stroke for an interquartile increase in mean daily pollutant concentration is shown. PM_{10} values include values interpolated for days when the actual value was missing and data were available for the preceding and following day. Abbreviations: ppm = parts per million, ppb = parts per billion, CI = confidence interval.

less), there was a 35% increase in the risk of a cerebrovascular event and an 83% increase in the risk of death from a cerebrovascular event [8]. The hazard ratio for the time to an acute cerebrovascular event, an indicator of relative risk, was reported as 1.35 (95% CI, 1.08–1.68). This prospective, observational study of postmenopausal women without a prior history of cardiovascular disease gains strength from the fact that the authors reviewed actual medical records, rather than relying on data from central databases. The restriction to this cohort of women loses some strength because it may not be possible to generalize the results to men or younger women. They did not find any association between other air pollutants, including sulfur dioxide, nitrogen dioxide, carbon monoxide, ozone, and stroke.

Two other studies, conducted under more restricted circumstances, have shown direct relationships between air pollutants and stroke. In a 2002 study of stroke mortality in Korea, Hong et al. reported significant and increasing risk for death due to ischemic but not hemorrhagic stroke as same-day particulate matter concentrations increased through four quartiles [9]. These authors also found significant temporal relationships between pollutants and stroke: same-day sulfur dioxide concentrations and ischemic stroke, a one-day lag between carbon monoxide and stroke, and a three-day lag between an ozone peak and stroke. In a more complex study of over 23,000 stroke admissions in Taiwan, Tsai et al. found that on days when the temperature was equal to or greater than 20 degrees centigrade (about 68 degrees Fahrenheit), there were significant positive associations between the concentrations of particles with a diameter of 10 microns or less, oxides of nitrogen and sulfur, carbon monoxide, and hospital admissions for ischemic strokes and intracerebral hemorrhages [10]. On cooler days, the correlation between carbon monoxide and ischemic stroke admissions was the only significant factor that persisted. When particles with a diameter of 10 μm and nitrogen oxide were considered together, there was a significant correlation between them and both types of stroke on warm days.

In summary, even though the increases in stroke admissions related to increases in pollutant levels are relatively small, on the order of 1% to 3%, the very large number of patients who have strokes each year makes particulate matter a risk factor that should be considered. These studies emphasize the importance of measures designed to minimize their concentration in the air.

Effects of Mercury

Coal contains trace amounts of mercury. Unless stringent control technologies are used to reduce mercury and other emissions, this mercury evaporates and is emitted into the environment when coal is burned. In 2005 coal-fired power plants were the most important anthropogenic source of mercury, as shown in table 3.3. More recent data were presented in the 2009 Toxics Release Inventory where point source air emissions of 6,277 pounds of mercury and 88,753 pounds of mercury compounds were reported [11]. Point sources of mercury are those that are not mobile, and consist primarily of power plants, cement kilns, and other boilers. Many of these burn coal.

History of Mercury Poisoning

Practically everyone has heard the expression, "mad as a hatter." This was most famously presented by Charles Lutwidge Dodgson writing as Lewis Carroll in his 1865 novel *Alice's Adventures in Wonderland* better known as *Alice in Wonderland*. Carroll refers to derangements observed among hat makers, attributed to their use of mercury to make felt. Two sentinel events are critical in the history of mercury poisoning and the development of regulatory criteria: the Minamata poisoning due to fish contaminated with methylmercury, and the Iraqi poisoning due to the ingestion of seeds treated with a mercury-containing fungicide.

In May 1956, several children were admitted to the city hospital in Minamata, Japan. Their symptoms were similar. They exhibited "intermittent lapses into [a] crazed mental state," epileptic seizures, episodic coma, and high fevers culminating in death [12]. The director of the Minamata city hospital, in cooperation with other medical centers and medical associations, began a systematic survey of the area. Subsequently he and his associates found that 17 other people living in fishing villages around Minamata City had died after identical illnesses. This was the beginning of what is now known as Minamata disease.

Data from this retrospective study showed that this outbreak had not started suddenly. The earlier cases had been observed, but the treating physicians had not recognized this constellation of signs and symptoms as a new disease. Initially these patients were thought to have infections. This hypothesis was discarded when more careful evaluation of the epidemiological data showed that all of the affected patients had eaten large amounts of fish from Minamata Bay.

It took several years of research by scientists at Kuamoto University to pinpoint the cause as a toxin in the fish and shellfish from Minamata Bay that came from the Chisso Minamata chemical complex. The investigation was complicated by several factors. The residents of Minamata were reluctant to blame this company because of its central importance in the community and the enormous effort that had been expended to persuade the company to build this factory. The company was not cooperative in the investigation and produced a report to refute the theory that heavy metals caused the disease even though a variety of toxic metals was identified in the sludge near the factory. The company also failed to reveal the fact that they used large amounts of mercury. Because of a reluctance to disbelieve the company report and the fact that mercury was expensive, the investigators expressed a doubt that discharges of mercury were likely. Finally, in July 1959, the researchers concluded that Minamata disease was indeed due to methylmercury poisoning. A riot ensued. Contrast this with the stoicism exhibited by the Japanese after the 2011 earthquake, tsunami, and Fukushima nuclear reactor failures.

There are approximately 3,000 officially recognized victims of Minamata disease, including just over 1,700 deaths. More than 10,000 people have been compensated. A memorial has been erected at the Minamata Disease Municipal Museum in memory of these victims. The Chisso factory continues to operate and is the largest employer in Minamata [13].

Over a decade later, in the fall of 1971, approximately 100,000 tons of wheat and barley seeds that had been treated with methylmercury, a fungicide, were distributed to various provinces in Iraq [14]. Despite the fact that the seeds had been dyed red, to indicate that they shouldn't be eaten due to the presence of the fungicide, these seeds found their way into the food chain after being milled into flour used to make bread. Apparently the dye that was intended to warn against consumption was removed easily by washing the kernels. The epidemic was intense, but short-lived. Medical records show that about 90% of those admitted to the hospital were affected in January and February of 1972. Although the epidemic was widespread, it affected rural populations almost exclusively. The highest frequency was found in Muthanna, where 7.6 out of every 1,000 inhabitants were affected. After a latent period of two to five weeks, a dose-dependent series of symptoms developed. The patients with the mildest cases complained of tingling in the toes and feet, and around the mouth. Unsteadiness while walking (ataxia) was common and ranged from mild ataxia to a severity that precluded the ability to walk. Visual disturbances varied from mild blurring to blindness. Slurred speech and hearing loss were present in some patients. Death was due to collapse of the central nervous system; cardiovascular, gastrointestinal, and renal disease were uncommon. At the time of the 1973 description of the epidemic 6,530 patients had been admitted to the hospital and 459 deaths were recorded.

Although the epidemic was a public health disaster, it is likely that only a relatively small amount of the total amount of mercury-treated wheat was eaten. Based on estimates of the total amount of mercury ingested (50–400 milligrams), affected individuals would have to have eaten between 6 and 50 kilograms of wheat.

Origin of Mercury Standards

Until relatively recently an extrapolation of the dose–response curve derived from the Minamata and Iraqi exposures served as the basis for determining the permissible daily intake of mercury [15]. Because of problems associated with this approach, alternate populations were examined in great detail.

The prospective study of Faroese children who had pre-natal exposure to mercury is perhaps the most important investigation of the effects of mercury on brain function [16]. Most important, instead of relying on data collected from highly exposed individuals, as in the case of the Iraqi and Minamata populations, this study involved individuals who had much lower mercury exposures. In addition it was done prospectively in a relatively homogeneous population, and brain function was evaluated using carefully selected neuropsychological and neurophysiological tests. Data that are collected prospectively, as in this study, are less likely to be influenced by bias due to factors that the investigator can't control, a problem that is common in studies that make use of preexisting data. Although prospective studies are expensive and time-consuming, the investigators have complete control of the study, a great advantage. Randomized clinical trials are the "gold standard" in terms of the strength of the evidence and conclusions. However, it would be unethical to conduct a randomized trial in which some of the participants were given methylmercury.

The Faroe islanders are a Nordic population of about 45,000 individuals. They are exposed to mercury as the result of eating meat from pilot whales, an important part of Faroese tradition [16]. These whales are at the top of the food chain. Squid are the whales' preferred food, but they will eat octopus, cuttlefish, herring, and other small fish if squid are not available. As they feed, the mercury in their prey becomes progressively more concentrated in the whales. The mercury from the whale meat, eaten by pregnant women, crosses the placenta and is concentrated further in the fetus [17]. Although the total number of Faroese is relatively small, the investigators were able to enroll 1,022 single-birth children (multiple births were excluded) over a 21-month period beginning in 1986 [16]. Hair samples were obtained from the mothers at the time of their child's birth along with blood samples from the umbilical cord. Cord blood mercury concentrations were as high as 350 micrograms per liter and 15% of the hair samples contained mercury in excess of 10 µg/g (micrograms per gram). Before this study, these levels would have been considered to be in the safe range.

Few Faroese leave the islands. Therefore, when the first large-scale follow-up study was done as these children reached 7 years of age (i.e., when they started school), a remarkable 90.3% of the surviving children were available for evaluation [16]. Seven children had died from causes not attributed to the effects of mercury. In addition to a comprehensive physical examination, two types of highly objective examinations were administered: electrophysiological tests and neuropsychological tests. The electrophysiological tests involve using mild stimuli, such as a click played into an ear or viewing a black-and-white checkerboard as the black and white squares alternate and recording the brain electrical activity associated with the stimuli. The neuropsychological tests involve performing specific tasks, such as inserting pegs into holes in a board, copying figures, and general tests of intelligence. These tests were chosen because they involve tasks that might be expected to be problematic because of pathological examinations of the brains of individuals exposed to mercury, such as the victims of the Minamata and Iraq poisonings. Careful statistical analyses were performed in an attempt to learn whether there were any relationships between mercury levels measured at birth and performance on these tests. The brainstem auditory evoked response was abnormal in these children. This test measures the length of time it takes a neural impulse to reach various brain structures after an auditory stimulus. The impulses traveled more slowly in the brains of these children as mercury exposures increased. The neuropsychological tests revealed evidence for impaired memory, attention, and language function as mercury levels increased. The authors of this report stressed the fact that the abnormalities they found were present at mercury exposure levels that were considered to be safe at that time.

Based on this study, and two other large-scale prospective studies of children exposed in utero to methylmercury in the Seychelles and New Zealand, the National Research Council recommended establishing a benchmark mercury (Hg) dose of 58 µg/L (micrograms per liter) in the cord blood of newborns [18]. The benchmark dose is the lower 95% confidence interval for an estimated dose that results in doubling the prevalence of children with test scores of neurodevelopment that are in the clinically impaired range. Based on these recommendations, the EPA then applied a tenfold safety factor and set the reference dose (RfD) for mercury at 5.8 µg/L Hg in cord blood. The reference dose is the maximum acceptable oral dose of a toxic substance. Fetal blood mercury levels are approximately 1.7 times those in the mother [17]. Based on the 1999 to 2000 National Health and Nutrition Examination Survey data, approximately 15.7% of all American women of childbearing age have blood mercury levels that would cause them to give birth to children with cord blood mercury levels of 5.8 µg/L Hg or more. Using these data and conservative estimates of blood mercury levels and their effect on intellectual performance, Trasande et al. estimated that between 316,588 and 631,233 children were born in the United States each year with blood mercury levels high enough to impair performance on neurodevelopmental tests [19]. These authors further concluded that this lifelong diminution in intelligence costs society $8.7 billion per year (range $2.2–$43.8 billion in year 2000 dollars). These cost estimates contrast sharply with others made by EPA officials during the George W. Bush administration that were as low as 10 million dollars [20].

Since mercury exists naturally in the environment, substantial amounts are released during volcanic eruptions, and mercury is recycled in the environment, opponents of stricter mercury controls attempt to downplay the importance of coal-related emissions. Nevertheless, minimizing or eliminating coal-related mercury emissions is the most important step that can be taken to prevent additional amounts of mercury from entering the environment and affecting health.

Summary

The neurological diseases attributed to coal-derived pollutants are varied. Many of the pollutants that affect the cardiovascular system leading to myocardial infarcts also affect the arteries that nourish the brain. Because of its unique biochemistry and physiology, the brain is highly susceptible to disruptions in its blood supply. Similar considerations make the brain more susceptible to the toxic effects of mercury. Although examples of overt mercury poisoning are rare, subtle effects on brain function cause significant impacts when large populations are examined.

References

1. NINDS. NINDS Stroke Information Page. Available at http://www.ninds.nih.gov/disorders/stroke/stroke htm. Accessed 2011.

2. Miniño AM, Xu J, Kochanek KD. Deaths: Preliminary Data for 2008. Centers for Disease Control. Atlanta: CDC, 2010.

3. Roger VL, Go AS, Lloyd-Jones DM, et al. Heart disease and stroke statistics—2011 update: a report from the American Heart Association. Circulation 2010;123(4):e18–209.

4. American Heart Assn Statistics Committee and Stroke Statistics Subcommittee. Heart disease and stroke statistics—2009 update. Circulation 2009;119: e21–181.

5. Dominici F, Peng RD, Bell ML, et al. Fine particulate air pollution and hospital admission for cardiovascular and respiratory diseases. JAMA 2006;295(10): 1127–34.

6. Blanchard C. Spatial and temporal characterization of particulate matter. In: McMurry PH, Shepherd MF, Vickery JF, eds., Particulate Matter Science for Policy Makers: A NARSTO Assessment. Cambridge: Cambridge University Press, 2004.

7. Wellenius GA, Schwartz J, Mittleman MA. Air pollution and hospital admissions for ischemic and hemorrhagic stroke among medicare beneficiaries. Stroke 2005;36(12):2549–53.

8. Miller KA, Siscovick DS, Sheppard L, et al. Long-term exposure to air pollution and incidence of cardiovascular events in women. N Engl J Med 2007;356(5): 447–58.

9. Hong YC, Lee JT, Kim H, Kwon HJ. Air pollution: a new risk factor in ischemic stroke mortality. Stroke 2002;33(9):2165–9.

10. Tsai SS, Goggins WB, Chiu HF, Yang CY. Evidence for an association between air pollution and daily stroke admissions in Kaohsiung, Taiwan. Stroke 2003; 34(11):2612–6.

11. US Environmental Protection Agency. TRI Explorer Chemical Report, Mercury and Mercury Compounds. Washington DC: GPO, 2011.

12. Ui J. Minamata disease. In: Ui J, ed., Industrial Pollution in Japan. Tokyo: United Nations University Press, 1992;ch.4.

13. Japan-guide.com. Minamata Disease Related sites. Available at http://www .japan-guide.com/e/e4527.html. Accessed 2011.

14. Bakir F, Damluji SF, min-Zaki L, et al. Methylmercury poisoning in Iraq. Science 1973;181(96):230–41.

15. Clarkson TW, Magos L. The toxicology of mercury and its chemical compounds. Crit Rev Toxicol 2006;36(8):609–62.

16. Grandjean P, Weihe P, White RF, et al. Cognitive deficit in 7-year-old children with prenatal exposure to methylmercury. Neurotoxicol Teratol 1997;19(6): 417–28.

17. Stern AH, Smith AE. An assessment of the cord blood:maternal blood methylmercury ratio: implications for risk assessment. Environ Health Perspect 2003;111(12):1465–70.

18. Committee on the Toxicological Effects of Mercury. Toxicological Effects of Methylmercury. Washington DC: National Research Council, National Academy Press, 2000.

19. Trasande L, Landrigan PJ, Schechter C. Public health and economic consequences of methyl mercury toxicity to the developing brain. Environ Health Perspect 2005;113(5):590–6.

20. Griffiths C, McGartland A, Miller M. A comparison of the monetized impact of IQ decrements from mercury emissions. Environ Health Perspect 2007;115(6): 841–7.

11

Health Effects on the Horizon

There are things we don't know we don't know.
—Donald Rumsfeld

It is hard to imagine that there was a time when chimneys belching clouds of black acrid smoke were seen as a sign of industrial prowess and economic vigor rather than a threat to health. At one time physicians appeared in advertisements touting the benefits of smoking cigarettes! Advances in knowledge change the way we look at our environment and how it impacts health. Where are the holes in our vision and what surprises are ahead? This chapter addresses that question using very early, sometimes provocative data to discuss what might emerge in the future and be placed on the growing list of health effects of the pollutants generated by burning coal.

Inflammation and oxidative stress are important in the pathogenesis of respiratory and cardiovascular diseases, as discussed in earlier chapters. Emerging data suggest that these two processes may affect the brain and metabolic pathways and play a role in the production of neurodegenerative diseases, particularly Alzheimer's disease (AD) and type 2 diabetes mellitus. If the early studies suggesting these links are upheld by adequately powered (i.e., large enough) more rigorous investigations that utilize contemporary epidemiological methods, there will be enormous public health implications.

Particulates and Neurodegenerative Disease

Alzheimer's disease is the most common form of dementia. In 2007 it was the sixth leading cause of death among Americans, claiming over

82,000 individuals [1]. It is the only disease among the top 10 for which there is no effective treatment to halt its progression. In addition to the human suffering by patients and caregivers, AD is also very expensive to manage. According to the Alzheimer's Disease Association more than 14 million Americans who are among the family or friends of AD patients provide over $200 billion in uncompensated care each year [2].

AD is characterized by the gradual loss of memory plus loss of function in at least one other aspect of mental capacity, such as judgment, language, or the ability to organize and execute tasks. These deficits must represent a decline from an earlier state and must be severe enough to interfere with social or occupational functions, or both. Although there are several drugs that are approved by the Food and Drug Administration for the treatment of AD, they all treat the symptoms of the disease but do nothing to halt the progression of its characteristic brain lesions.

Animal data suggest strongly that very small particles, those with a diameter of 2.5 microns or less, cross the membranes of the nose and enter the brain via the olfactory nerve [3,4]. This is the nerve that mediates the sense of smell. The endings of this nerve begin in the nose and travel through many small holes in the skull (the cribiform plate) and enter the brain, forming the olfactory nerve. This nerve and its connections in the brain form part of what is known as the limbic system, which plays a key role in the mediation of memory, emotions, and other basic neural functions. The hippocampus is an important part of the limbic system. Pathological changes in the hippocampus and other parts of the brain, specifically deposition of a protein known as β-amyloid, the loss of neurons and synapses (connections between neurons), and the presence of neurofibrillary tangles, are the neuropathological hallmarks of Alzheimer's disease.

Doty reviewed what he calls the "olfactory vector hypothesis of neurodegenerative disease," namely the hypothesis that environmental toxicants or other agents enter the brain via the olfactory nerve leading to the production of diseases such as AD and idiopathic Parkinson's disease (PD) [5]. In support of the hypothesis, he notes that numerous studies have shown that tests of olfactory function are abnormal in as many as 90% of all patients with early stages of AD and PD. Pathological changes that are characteristic of both PD and AD are found in the olfactory pathways, providing further support of the hypothesis [6,7]. Hawkes et al. expanded this theory by proposing a "dual-hit" hypothesis, suggesting that offending agents may gain entry to the nervous system via the nose and collections of nerves (neural plexuses) in the stomach [8]. From the stomach, the agents could ascend into the brain in the fibers that form the vagus nerve.

Some of the most intriguing and potentially important data linking coal-derived air pollution and neurological disease have come from studies comparing brains of dogs and humans living in highly polluted versus nonpolluted cities in Mexico. In Mexico City, airborne particulate matter levels (as well as those of other pollutants) regularly exceed US air quality standards. This is shown in figure 3.3. In a study of the brains of 26 dogs from Mexico City, the concentration of the 42 amino acid form of β-amyloid and the expression of cyclooxygenase were increased compared to levels in the brains of dogs from regions with low pollutant levels [3]. These findings may be relevant because β-amyloid deposits are characteristic of AD neuropathology in humans and cyclooxygenase is an enzyme that is found in inflammatory conditions. Similar results were found in a second study of 10 human brains (age 51.2 ± 4.9 years, standard deviation) from individuals who had lived in high-pollution areas compared to 9 brains (age 58.1 ± 4.6 years, standard deviation) from those who had lived in low-pollution areas [9]. None of the brains in that study were from patients with known neurological disease or cognitive deficits. The authors suggest that "exposure to urban air pollution may cause brain inflammation and accelerate the accumulation of [β-amyloid$_{42}$], a putative mediator of neurodegeneration and AD pathogenesis." These data are supported by the results of animal experiments in which several strains of transgenic mice were exposed to particulate matter [10].

To further test the olfactory hypothesis, Calderón-Carcidueñas et al. found significant reductions in a test of smell in a cohort of 65 individuals from Mexico City, the pollution-exposed group, when compared to 25 controls, drawn from a less polluted area [11]. In that same publication they reported the results of pathological examinations of the olfactory bulb (a part of the olfactory nerve and the limbic system) in 35 autopsy specimens from residents of Mexico City and 9 controls. The average age was 20.8 years. Ultrafine particles were found in the olfactory bulb of 2 of the specimens from Mexico City residents but none of the controls. Twenty-nine of the Mexico City olfactory bulbs contained β-amyloid whereas none of the control bulbs contained this protein. The authors concluded that air pollution may play a key role in the development of neurodegenerative diseases.

Although there are characteristics of particulate matter that define its origin, at present the EPA treats all particulates similarly (i.e., the Agency does not distinguish effects of particulates that appear to be related to vehicular traffic from those originating from coal and other sources). Thus reports that focus on diesel exhaust particles may be completely relevant to particles originating from fossil fuel combustion.

Two reports relating traffic-caused pollution to cognitive function are worthy of consideration. Ranfit et al. studied a group of 399 women between the ages of 68 and 79 years of age who had lived for 20 or more years at the same address [12]. Particulate exposure was highest among those who lived near busy highways. Mental function was evaluated using a neuropsychological test battery and smell was evaluated with an odor identification test. They report a dose–response function linking decrements in mental function and increases in particulate concentration that survived statistical adjustments for potential confounders. Specifically, the results of tests for mild cognitive impairment, a risk factor for Alzheimer's disease, were significantly poorer if the participant lived within 50 meters of a highway with a traffic density of 10,000 cars or more per day. A study in which the investigators sought a relationship between nitrogen dioxide levels and cognitive function in 210 children who were 4 years old found a suggestive deleterious effect that was not statistically significant [13].

The possible link between airborne pollutants and neurodegenerative diseases is still very speculative. However, it may ultimately be extremely important because of the very large and growing number of patients with Alzheimer's disease and the financial and societal impacts of this terrible illness.

Air Pollution and Diabetes Mellitus

Diabetes mellitus is the seventh leading cause of death in Americans [1]. According to the CDC, 70,610 Americans died because of diabetes in 2008. Diabetes is one of the most debilitating of all chronic diseases, particularly when it is poorly controlled. Diabetes may lead to the development of hypertension and is a risk factor for strokes and heart attacks. Patients with advanced poorly controlled diabetes may develop kidney failure, requiring dialysis or transplantation, and may suffer repeated amputations or become blind.

Diabetes is one of the most common and most frequently undiagnosed chronic diseases. A recent estimate of the worldwide prevalence of diabetes predicts that the prevalence of this condition will continue to increase, particularly in underdeveloped countries [14]. The authors of this report drew on data from 91 countries to estimate the prevalence of diabetes in 2010 and 2030 in 216 nations. They estimated that among adults, aged 20 to 79, the prevalence of diabetes in 2010 would be 6.4%, affecting 285 million individuals. By 2030, the prevalence in developing nations would increase by 69%, and in developed nations the prevalence is expected to increase by 20%. By 2030, they estimated that 7.7% of adults, or 439 million individuals, will be diabetic worldwide.

Diabetes is a disease that is characterized by an elevation of the amount of glucose in the blood. There are two main types: type 1 diabetes, sometimes called juvenile diabetes, is a form of the disease in which the beta cells of the pancreas fail, making patients dependent on insulin. In type 2 diabetes, formerly called adult-onset diabetes, the body fails to respond appropriately to the insulin that is produced by the pancreas. Diabetes in these patients is treated with a combination of medications, diet, weight loss, and exercise. Some women develop diabetes during pregnancy, a condition called gestational diabetes. The symptoms of diabetes include abnormal thirst, frequent urination, increased hunger, weight loss, and fatigue. Diabetes is diagnosed by measuring the amount of glucose in the blood.

The rising prevalence of diabetes mellitus, particularly type 2 diabetes, is a cause for national concern. Many regard this as an epidemic that is out of control. In 2002, I made the serendipitous observation that there was a statistically significant correlation between the by-state prevalence of diabetes and the by-state emission of pollutants discharged into the air as reported via the Toxics Release Inventory ($r = 0.54, p = 0.000054$) [15]. I speculated that airborne dioxins might be the cause, as others had implicated this class of chemicals as a risk factor for the development of diabetes [16].

In 2005 Brook postulated several mechanisms by which the inhalation of particulates might lead to the development of insulin resistance, a pre-diabetic condition characterized by an inability to respond normally to insulin [17]. These mechanisms begin with the inhalation of small particles that stimulate pulmonary inflammation and generate reactive oxygen species and oxidative stress. One pathway is suggested to operate via the brain and the autonomic nervous system, leading to stimulation of the adrenal gland with subsequent increases in epinephrine and cortisol. Both of these hormones affect glucose metabolism. A second and perhaps concurrent pathway was suggested that might operate more directly on blood vessels. In concert with autonomic nerves, this could lead to vasoconstriction and insulin resistance.

This hypothetical mechanism gained substantial credence with the publication of a paper linking nitrogen dioxide exposure to diabetes in women [18]. In this study the authors evaluated the records of over 7,500 patients attending clinics for respiratory diseases in Toronto and Hamilton, Ontario. Geographic information systems techniques were used to estimate nitrogen dioxide exposures from networks of air sampling devices in each of the cities. The diagnosis of diabetes was obtained from a provincial medical database. There was a positive relationship between nitrogen dioxide exposure and diabetes in both cities, after controlling for several potential confounding factors such as age and body mass index. When the data from both cities were combined there was about a 17% increase in the odds for diabetes in women between quartiles of exposure (quartiles divide the population into four sections, based on the nitrogen dioxide exposure). Although the authors of this report used nitrogen dioxide levels as a surrogate marker for traffic-related air pollutants, substantial amounts of this pollutant are formed when coal is burned. The authors conclude that their results "suggest that common air pollutants are associated with [diabetes mellitus]."

A second study linking exposures to particulates and nitrogen dioxide with diabetes was the result of 16 years of following 1,775 nondiabetic German women who were between the ages of 54 and 55 at the time of enrollment [19]. Using questionnaires that specified physician-diagnosed type 2 diabetes and NO_2 and particulate matter concentrations at locations associated with each patient, the investigators found significant relationships between these two pollutants and the risk of developing diabetes. The relationship was stronger for NO_2 than for particulates, suggesting to them that traffic may be more important than other sources of pollution. They also measured complement C3c, a marker for low-level inflammation, and found that those with the highest levels of the marker had the highest risk for developing diabetes.

Finally, a US study of the relationship between fine particles and the prevalence of diagnosed diabetes found a significant relationship between these variables [20]. In counties where the fine particle concentration was within EPA air quality standards, the prevalence of diabetes among those with the highest exposure was more than 20% higher than the prevalence among those with the lowest exposures. They estimated the risk for developing diabetes was on the order of 1% for an increase in the concentration of fine particles of 10 $\mu g/m^3$.

The evidence linking diabetes with air pollutants that are derived, in part, from burning coal is becoming increasingly convincing. As is the case with neurodegenerative disorders, if this relationship gains strength, the public health implications will be enormous. The observation that the risk for the development of diabetes increases even within the current air quality standards for particulates is especially telling. As is the case with other disease conditions, there may well not be a limit below which one can be free of risk. A re-evaluation of the standard is in order, as mandated by the Clean Air Act. The failure of the EPA to begin this process is the subject of legal action initiated by Earthjustice.

Are Diabetes and Alzheimer's Disease Linked?

The March 2009 issue of the Archives of Neurology was devoted to a series of papers discussing the possible links between metabolic disorders, such as diabetes and the metabolic syndrome, with Alzheimer's disease and vascular dementia. Metabolic syndrome is a disorder with several defining characteristics, including abnormal glucose metabolism, elevated blood triglycerides, low blood levels of high-density lipoproteins, hypertension, and central obesity, meaning an elevated waist-to-hip circumference ratio. Metabolic syndrome is an important risk factor for the development of diabetes.

In a summary article in that issue, Craft discusses evidence that insulin resistance is related to Alzheimer's disease [21]. Insulin resistance is a diabetes risk factor and is characterized in part by elevated blood insulin levels and a lack of responsiveness of various organs and tissues to the effects of insulin. Although the putative pathophysiological mechanisms that link these two diseases are complex, and beyond the scope of this review, inflammation and oxidative stress appear prominently. As discussed throughout this book, these two mechanisms are intimately related to exposure to various air pollutants, particularly fine particles.

Alzheimer's disease and diabetes are two of the most common and costly chronic diseases in the United States. The prevalence of both is increasing dramatically. It seems quite possible that the list of the many adverse health effects of burning coal will expand beyond the obvious (i.e., respiratory diseases), to include a substantially larger portion of the population. Eventually, burning coal to produce electricity may be seen as an addictive, unacceptable universal threat to health, a modern-day equivalent to smoking cigarettes.

References

1. Miniño AM, Xu J, Kochanek KD. Deaths: Preliminary Data for 2008. Atlanta: CDC, 2010.

2. Alzheimer's Disease Association. 2011 Alzheimer's Disease Facts and Figures. Available at http://www.alz.org/documents_custom/2011_Facts_Figures_Fact _Sheet pdf. Accessed 2011.

3. Calderón-Garcidueñas L, Maronpot RR, Torres-Jardon R, et al. DNA damage in nasal and brain tissues of canines exposed to air pollutants is associated with evidence of chronic brain inflammation and neurodegeneration. Toxicol Pathol 2003;31(5):524–38.

4. Oberdorster G, Sharp Z, Atudorei V, et al. Translocation of inhaled ultrafine particles to the brain. Inhal Toxicol 2004;16(6–7):437–45.

5. Doty RL. The olfactory vector hypothesis of neurodegenerative disease: is it viable? Ann Neurol 2008;63(1):7–15.

6. Ohm TG, Braak H. Olfactory bulb changes in Alzheimer's disease. Acta Neuropathol 1987;73(4):365–9.

7. Pearce RK, Hawkes CH, Daniel SE. The anterior olfactory nucleus in Parkinson's disease. Mov Disord 1995;10(3):283–7.

8. Hawkes CH, Del TK, Braak H. Parkinson's disease: a dual-hit hypothesis. Neuropathol Appl Neurobiol 2007;33(6):599–614.

9. Calderón-Garcidueñas L, Reed W, Maronpot RR, et al. Brain inflammation and Alzheimer's-like pathology in individuals exposed to severe air pollution. Toxicol Pathol 2004;32(6):650–8.

10. Peters A, Veronesi B, Calderón-Garcidueñas L, et al. Translocation and potential neurological effects of fine and ultrafine particles a critical update. Part Fiber Toxicol 2006;3:13.

11. Calderón-Garcidueñas L, Franco-Lira M, Henriquez-Roldan C, et al. Urban air pollution: influences on olfactory function and pathology in exposed children and young adults. Exp Toxicol Pathol 2010;62(1):91–102.

12. Ranft U, Schikowski T, Sugiri D, Krutmann J, Krämer U. Long-term exposure to traffic-related particulate matter impairs cognitive function in the elderly. Environ Res 2009;109(8):1004–11.

13. Freire C, Ramos R, Puertas R, et al. Association of traffic-related air pollution with cognitive development in children. J Epidemiol Community Health 2010;64(3):223–8.

14. Shaw JE, Sicree RA, Zimmet PZ. Global estimates of the prevalence of diabetes for 2010 and 2030. Diabetes Res Clin Pract 2010;87(1):4–14.

15. Lockwood AH. Diabetes and air pollution. Diabetes Care 2002;25:1487–8.

16. Michalek JE, Akhtar FZ, Kiel JL. Serum dioxin, insulin, fasting glucose, and sex hormone-binding globulin in veterans of Operation Ranch Hand. J Clin Endocrinol Metab 1999;84(5):1540–3.

17. Brook RD. You are what you breathe: evidence linking air pollution and blood pressure. Curr Hypertens Rep 2005;7(6):427–34.

18. Brook RD, Jerrett M, Brook JR, Bard RL, Finkelstein MM. The relationship between diabetes mellitus and traffic-related air pollution. JOEM 2008;50(1):32–8.

19. Krämer U, Herder C, Sugiri D, et al. Traffic-related air pollution and incident type 2 diabetes: results from the SALIA cohort study. Environ Health Perspect 2010;118(9):1273–9.

20. Pearson JF, Bachireddy C, Shyamprasad S, Goldfine AB, Brownstein JS. Association between fine particulate matter and diabetes prevalence in the U.S. Diabetes Care 2010;33(10):2196–2201.

21. Craft S. The role of metabolic disorders in Alzheimer disease and vascular dementia: two roads converged. Arch Neurol 2009;66(3):300–5.

12

Coal, Global Warming, and Health

If I have seen further, it is by standing on the shoulders of giants.
—Isaac Newton

Charles David Keeling, like many scientists, was a man driven by the compulsion to be right. In the 1950s he focused his talents on measuring the very low levels of CO_2 in the atmosphere with great accuracy and precision. After perfecting his technique, he established a sampling site at the Mauna Loa Observatory in Hawaii. Each hour the successors to his early device churn out the current CO_2 concentration in the atmosphere. The time concentration graph, now known worldwide as the Keeling Curve, is shown in figure 12.1. It is recognized as one of the major accomplishments in all of science. As a result Keeling was awarded the National Medal of Science by President George W. Bush in 2002.

The Keeling Curve has two major features: predictable seasonal fluctuations and an otherwise relentless rise. When Keeling began his studies, the concentration of CO_2 was just over 310 ppm (i.e., 310 molecules of CO_2 per million molecules of air). By the time this book goes to press, the level is likely to be close to 400 ppm. The seasonal dip in the CO_2 concentration is due to springtime tree growth and leaf sprouting in the forests of the Northern Hemisphere. These processes both trap CO_2. The rise that occurs in the fall and winter is due to the release of carbon as leaves fall. Both of these forces are superimposed on the steady addition of CO_2 due to burning fossil fuels.

The near-continuous monitoring of atmospheric CO_2 levels was critical to establishing the validity and importance of the data. Without this uninterrupted record, the seasonal changes would have been missed and differentiating a real change from an artifact would have been much

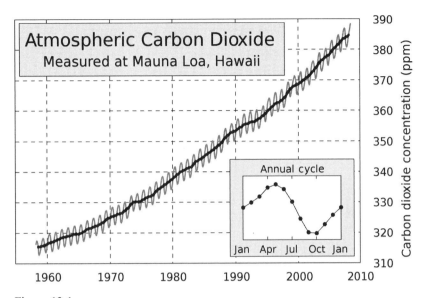

Figure 12.1
The Keeling Curve. This is the record of the concentration of carbon dioxide in the atmosphere at the Mauna Loa Observatory from the late 1950s to the present. It shows a steady increase in the concentration of this important greenhouse gas and the annual fluctuations. The decreases in the concentration during the Northern Hemisphere's spring are due to plant growth and carbon dioxide sequestration. In the fall, this ends, and carbon dioxide levels rise during the winter months. Source: Wikimedia Commons, not copyrighted.

more difficult. The steady rise in the concentration of CO_2 and its high correlation with fossil fuel combustion were seminal findings that led to the development of the currently accepted hypothesis that human activity is causing global warming.

Now there are numerous sites around the world that measure the atmospheric concentration of the important long-lived greenhouse gases. Figure 12.2 illustrates the increases in CO_2, methane, and nitrous oxide that have occurred over the course of the past two thousand years.

The work of Keeling, and the huge number of scientists who saw further because of him, led to the award of the 2007 Nobel Peace Prize to the Intergovernmental Panel on Climate Change and Albert Gore Jr. "for their efforts to build up and disseminate greater knowledge about man-made climate change, and to lay the foundations for the measures that are needed to counteract such change." Confronting climate change is arguably the most important challenge facing us today.

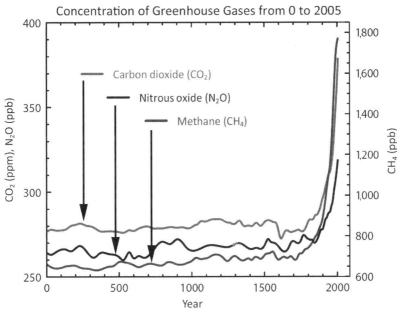

Figure 12.2

Atmospheric concentrations of important long-lived greenhouse gases over the last 2000 years. Increases since about 1750 are attributed to human activities in the industrial era. Concentration units are parts per million (ppm) or parts per billion (ppb), indicating the number of molecules of the greenhouse gases per million or billion air molecules, respectively, in an atmospheric sample. Source: Intergovernmental Panel on Climate Change, reproduced in accord with their copyright requirements [2].

The Greenhouse Effect and Global Warming

The greenhouse effect is a term used to describe the process by which some of the energy from the sun is captured and retained by the earth in the form of heat. Most solar energy reaches us on earth in the form of light with a relatively short wavelength. Some of this light energy is reflected back into space, but almost half warms the earth's surface and our atmosphere. Some of the energy that is deposited at the earth's surface is radiated back into space in the form of longer wavelength infrared radiation or heat. Some of this infrared energy is absorbed by gases in the atmosphere and re-radiated in all directions, including back to the earth. It is this re-radiation back toward the earth that is popularly known as the greenhouse effect. Ironically, by this definition, greenhouses do not exhibit a greenhouse effect. Heat is trapped in greenhouses by the physical barrier imposed by the glass that limits mixing of warm air inside the greenhouse with cooler outside air.

Atmospheric gases vary substantially in their ability to absorb the infrared radiation that comes from the earth's surface. Gases that absorb a large fraction of this infrared radiation make the greatest contribution to the greenhouse effect and climate change. There are three factors that determine the importance of a given gas to the production of global warming: the length of time a gas resides in the atmosphere, its concentration, and radiative forcing. Radiative forcing is a complex concept that is not easily explained in plain language. One of the simplest definitions, published by the Intergovernmental Panel on Climate Change, is that it is "a measure of the influence a factor has in altering the balance of incoming and outgoing energy in the earth–atmosphere system and is an index of the importance of the factor as a potential climate change mechanism." Gases with high radiative forcing have the largest effect on climate change.

Radiative forcing values refer to changes relative to pre-industrial conditions defined in 1750 and are expressed in units of watts per square meter (W/m^2) [1]. Radiative efficiency is related to radiative forcing and includes a concentration term that allows one to make more meaningful comparisons between various gases. Radiative efficiency units are measured in watts per square meter at a given concentration. Its concentration units are expressed as parts per billion ($Wm^{-2}ppb^{-1}$). Radiative forcing and related terms applicable to greenhouse gases are fairly complex and not easily understood on an intuitive basis.

This leads to a final, important term that synthesizes these various factors: the global warming potential (GWP). Gases with a high GWP make greater contributions to global warming than gases with a low GWP. A complete, formal definition of GWP is a bit complicated, but it is influenced primarily by the radiative forcing value for the gas and the length of time the gas resides in the atmosphere. The time of residence varies substantially among the greenhouse gases (for a complete discussion, see the Intergovernmental Panel on Climate Change reports [1,2]). Thus global warming potential values, by definition, include a time factor. By convention, all GWP values are referenced to CO_2, which is assigned a value of unity for all times considered. All of this might be made clearer by examining the specific examples shown in table 12.1.

Table 12.1

Contribution of global warming gases attributable to human activity, omitting halogenated compounds [1,2]

Gas	Atmospheric lifetime (years)	Radiative forcing, 2005 (W m²)	Radiative efficiency (W m⁻² ppb⁻¹)	Radiative efficiency ratio compared to CO_2	Concentration in atmosphere (2005)	20 Year GWP	100 Year GWP	500 Year GWP
Carbon dioxide	Very long, see below[a]	1.66	1.4×10^{-5}	1	379 ± 0.65 ppm	1	1	1
Methane	12	0.48	3.7×10^{-4}	0.264	$1,774 \pm 1.8$ ppb	72	25	7.6
Nitrous oxide	114	0.1.6	3.03×10^{-3}	0.0046	319 ± 0.12 ppb	289	298	153

Note: Abbreviations: W = watts, m = meter, ppb = parts per billion, GWP = global warming potential.
a. The atmospheric lifetime for carbon dioxide is difficult to determine with great precision because of the complexity of the earth's carbon budget [1,2].

The global warming potential at all times for methane and nitrous oxide is greater than that of carbon dioxide because of the higher radiative forcing values for these two gases. However, since the concentration of CO_2 in the atmosphere is so much higher than the concentration of methane or nitrous oxide, the effect of atmospheric CO_2 is very high, as shown by the radiative efficiency ratio. This ratio is computed by dividing the radiative efficiency for a given gas by the efficiency for CO_2.

Since the initial groundbreaking observations of Keeling, there has been an explosion in climate research. We now know a great deal more about the causes of global warming. Now it is well understood that the climate of the earth is governed by a complex interaction of factors that work to increase and decrease global temperatures. CO_2, methane (CH_4), nitrous oxide (N_2O), halocarbons (organic molecules that contain chlorine, fluorine, or some other halogen), ozone in the lowest layer of the atmosphere (the troposphere), stratospheric water vapor, contrails produced by aircraft, and black carbon deposits on snow all act to promote warming. Stratospheric ozone, net increases in the reflectivity of the earth (the earth's albedo) due to the direct effects of atmospheric aerosols and aerosols in clouds, all act to promote cooling [2]. Of these factors, CO_2 and CH_4 are the most directly related to coal production and use and make major impacts on climate change, in the direction of warming the earth.

The earth will get warmer as the result of the changes in the atmosphere that have already taken place and those that are expected to occur in the future. In the Intergovernmental Panel on Climate Change Fourth Assessment Report, the panels reviewed six different models that have been used to predict the effects of changes in the atmosphere on global and regional temperatures and on sea level. Although the details differ, all of the reports predict substantial temperature increases.

Because of the long-lasting effects of global warming gases, the amount of the gases already in the atmosphere, and the lag between emission and final effects, global temperatures are expected to increase substantially by the end of this century. The smallest impact on climate predicts a temperature increase of 0.6 degrees centigrade even if no additional greenhouse gases are released (likely range = 0.3–0.9 degrees) [1]. Using 1980 to 1999 data as a baseline for comparison, the best estimate for the increase in global temperatures in this century ranges from 1.8 to 4.0

degrees centigrade, depending on which of the six models the group considered [1]. The likely *range* across the six studies (i.e., the lowest low and the highest high estimates) included in the report varied from a 0.3 degree increase to a high of 6.4 degrees.

The various models make regional as well as overall predictions. Warming is expected to be greatest at high northern latitudes, particularly over land masses, with less warming over southern oceans near Antarctica and over northern ocean areas. This is expected to lead to reductions in the land area covered by snow that will have secondary effects on the earth's albedo, or the ability to reflect light energy. Since snow reflects more heat energy than bare soil, this darkening of the earth will reduce the amount of heat reflected back into space. In areas of tundra, warmer land masses that are not covered by snow will lead to thawing of the permafrost. This, in turn, may lead to the release of methane that is currently trapped in the form of frozen methane hydrates. These hydrates, or more properly clathrate hydrates, contain large amounts of methane trapped in the crystalline structure of ice.

Ocean levels will rise [1]. The six models predict an increase in sea level that ranges between 0.2 and 0.6 meters. Some of this increase is due to melting snow and ice and some due to an expansion of the water in the oceans as they become warmer. In the years beyond 2100, thinning, or, in a worst case scenario, complete loss of the Greenland ice cap, could lead to an increase in sea levels of as much as 7 meters.

Carbon Dioxide

Carbon dioxide is the single most important global warming gas that is produced as the result of human activity. It is formed chiefly by burning fossil fuels, making cement, and flaring or burning of methane during the production of natural gas [2]. The United Nations Millennium Development Goals Indicators lists world-wide CO_2 emissions of 32.25 billion tons in 2007 [3]. Selected emissions by country were reported as follows: China, 71.92 billion tons; United States, 6.422 billion tons; India, 1.774 billion tons; the Russian Federation, 1.691 billion tons; Japan, 1.380 billion tons; France, 409 million tons; and Germany, 867 million tons. The best estimates suggest strongly that these emissions will continue to increase into the foreseeable future.

Global warming deniers question whether human activity is really responsible for the increases in atmospheric CO_2. Could this be a natural phenomenon? Science has ready answers [2]. First, the increases in atmospheric CO_2 are highly correlated with increases in the consumption of fossil fuels. As all statisticians know, correlation does not prove a cause-and-effect relationship. The principles of evidence-based medicine, as summarized in chapter 7, hold that a cause-and-effect relationship becomes more likely when several conditions are met: (1) the relationship is plausible, (2) the strength of the relationship is high, meaning the correlation is statistically significant, (3) there is consistency among studies, and (4) there is evidence from other investigations that supports a cause-and-effect relationship. All of these conditions are met in a highly convincing manner, and burning coal is the central culprit.

The evidence from other studies that supports the global warming hypothesis provides further insights into the workings of science. As one burns fossil fuels, the carbon combines with oxygen in the atmosphere to form CO_2. Therefore, as taught in introductory chemistry classes, there should be a concomitant decrease in the amount of oxygen in the atmosphere. This was a challenging problem to approach until appropriate methods were developed. Part of the problem centered on the difficult task of making accurate measurements of a very small difference in oxygen concentration amid the very high oxygen concentration in the earth's atmosphere. Interestingly it was Charles Keeling's son, Ralph, who was instrumental in developing a solution to this problem. In 1992 he and his colleague showed that the concentration of oxygen in the atmosphere has decreased and varies seasonally, in synchrony with the fluctuations in the concentration of CO_2 [4]. This relationship between O_2 and CO_2 is made more complex by the facts that photosynthesis produces O_2 and oceans sequester large amounts of CO_2. Thus there is not a simple one-to-one relationship between these two gases.

Perhaps the most powerful evidence linking the increases in atmospheric CO_2 to burning fossil fuels came from an analysis of the isotopes of carbon in atmospheric CO_2. Carbon has two principle isotopic forms: ^{13}C, which accounts for about 1% of atmospheric carbon, and ^{12}C, which accounts for about 99%. Both are nonradioactive, stable isotopes. Biological reactions discriminate against ^{13}C, so the carbon in plants contains lower amounts of this isotopic form of carbon than one might expect from the ratio that occurs naturally in the atmosphere [5]. It follows that CO_2 formed by burning plant-derived coals will also contain less ^{13}C than CO_2 derived from other sources. Thus it is possible to construct a "fingerprint" of atmospheric carbon by measuring this isotopic ratio and to deduce the origin of atmospheric CO_2.

The emissions associated with burning coal and other fossil fuels have ratios of $^{13}C/^{12}C$ that are lower than the ratio in the atmosphere. Thus, when carbon dioxide produced by burning coal enters the atmosphere, the atmospheric $^{13}C/^{12}C$ ratio should fall. This is exactly what has been observed [6]. The converging lines of evidence clearly establish burning fossil fuels as the source of the seemingly relentless increase in the atmospheric CO_2 concentration.

Humans burn many forms of carbon. Each has a defined amount of CO_2 that is produced per unit of heat produced. Unfortunately, coals produce more CO_2 per unit of heat generated than most other sources. This information is available from the US Energy Information Administration [7]. Anthracite coal, because of its high carbon and low water content, is the worst offender. It produces about 104 kilograms of CO_2 per million British thermal units. Bituminous coal produces about 93 kg of CO_2 per million British thermal units. Natural gas, however, typically produces about 53 kg of CO_2 per million British thermal units. Since the amount of heat determines the amount of steam and thus electricity that is produced, burning natural gas produces about 57% as much CO_2 as is produced by coal. Although all CO_2 released into the atmosphere contributes to global warming, natural gas is less of an offender than coal.

Methane

Next to CO_2, methane makes the largest contribution to global warming of all of the long-lived global warming gases [2,4]. As is the case for CO_2, atmospheric methane concentrations have increased substantially since the beginning of the industrial era. According to the Fourth Working Group of the Intergovernmental Panel on Climate Change, methane levels in bubbles trapped in ice show that levels ranged from about 400 ppb (parts per billion) during periods of glaciation to 700 ppb during interglacial periods [2,4]. Current methane levels are approximately 1,770 ppb, measured at multiple sites in both the Northern and Southern hemispheres. These levels are higher than any recorded for at least

650,000 years. Atmospheric methane levels may be plateauing, with little if any change between the years 1999 to 2005, the only encouraging aspect of these data. However, if the massive amount of methane that lies beneath the frozen tundra in arctic regions is released due to temperature increases, the concentration of methane in the atmosphere may soar.

The Fourth Assessment of the Intergovernmental Panel on Climate Change lists the following anthropogenic sources of atmospheric methane, in descending order of importance: cows and other ruminants; rice agriculture; landfills and waste treatment; natural gas, oil, and other industrial processes; coal mining; and others including biomass burning. Natural sources include termites, oceanic sources, hydrates, wild animals, wetlands, geological sources, and wildfires. Methane is removed from the atmosphere by a variety of processes. Methane reacts with OH free radicals in the atmosphere, ultimately forming CO_2 and water. This chemical reaction is critical and the most important of the mechanisms that remove methane from the atmosphere.

Global Warming and Health

The 1948 Preamble to the Constitution of the World Health Organization defined health: "Health is a state of complete physical, mental and social well-being and not merely the absence of disease or infirmity." The threats to health posed by global warming may be unique in their scope and scale. The complexity of these threats to health is illustrated in figure 12.3 [8].

The effects of global climate changes on health are not an abstract concept dealing with the future. The World Health Organization reports that malnutrition, diarrheal diseases, malaria, floods, and cardiovascular disease attributable to global warming were responsible for an increase of 166,000 deaths per year in 2000 compared to baseline data recorded between 1961 and 1990 [9]. In addition these diseases caused an increase of approximately 5.5 million disability-adjusted life years (DALYs) per year in that same time period, as shown in figure 12.4 [9].

Future changes in our climate will have direct effects on environmental and social conditions due to increases in temperature, changes in precipitation amounts and patterns, rising sea level, and the likelihood

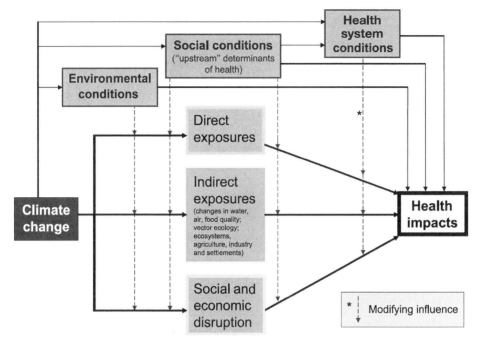

Figure 12.3
Schematic diagram by which climate change affects health, and concurrent direct-acting and modifying (conditioning) influences of environmental, social, and health-system factors. Source: Figure 8.1 from Confalonieri et al. reproduced in accord with their copyright requirements [8].

that there will be more extreme weather events. Less direct consequences are likely to affect air quality, access to safe drinking water, disruptions to the food supply associated with effects on agricultural systems, ecosystems, and, at the bottom line, the overall economy. Simply stated, every element in the health pathways illustrated in figure 12.3 affects every other element.

Effects of Increased Temperature
An increase in temperatures throughout the world is perhaps the most tangible effect of global warming. This may also be the simplest to understand. Before proceeding, it is particularly important to draw a clear line separating climate and weather. Weather is what is happening right now, right outside your window. The *Oxford English Dictionary*

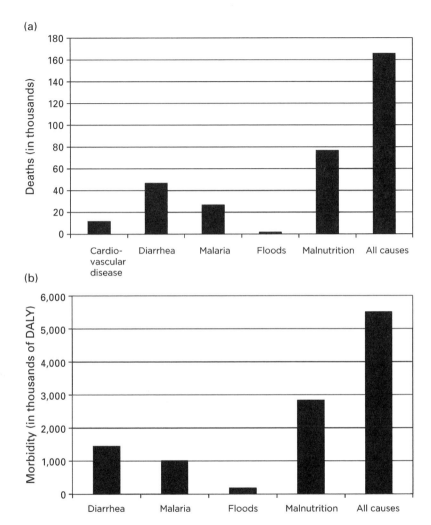

Figure 12.4
Morbidity and mortality associated with global warming. (a) Deaths in thousands due to global warming in 2000 compared to baseline data recorded between 1961 and 1990. (b) Morbidity, expressed in thousands of DALYs (disability adjusted life years), a measure of disease burden that combines years of life lost with years of life lived with illness or disability. Source: Global Climate Change [9].

has an elegant definition, "The condition of the atmosphere (at a given place and time) with respect to heat or cold, quantity of sunshine, presence or absence of rain, hail, snow, thunder, fog, etc., violence or gentleness of the winds" [10]. Climate refers to these elements averaged over a long period of time. As climate changes, it is likely that there will be fewer gentle winds and more that are violent. As the average temperature increases over the period of years or decades due to climate change, there will be periods of time when the weather is hot or cold, when it rains, or when the sun shines, and so forth.

During their deliberations in 2007, the Intergovernmental Panel on Climate Change Working Group I reviewed data from temperature sensors throughout the globe, satellite data, the temperature of the ocean, and so forth. They found that eleven of the twelve hottest years dating back to 1850 occurred in the twelve year interval between 1995 and 2006. On the basis of these data, the Working Group concluded that it was "virtually certain" that during the twenty-first century it would be warmer with fewer cold days and nights and more frequent hot days and nights.

A heat wave exists when the maximum daily temperature exceeds the average maximum daily temperature by 5 degrees centigrade (9 degrees Fahrenheit) for five days or more, using the 1961 to 1990 data as a reference [11]. By this definition, historical meteorological data have shown that heat waves were more common in the second half of the twentieth century than in the first half [11].

The most infamous of recent heat waves occurred over large portions of Europe during the summer of 2003. In parts of Italy, the temperatures were 16 degrees Fahrenheit higher than in the prior year, temperatures over 100 degrees were recorded for the first time in London, and records were broken throughout many parts of Europe [12]. Hospitals and undertakers were overwhelmed by the number of those with heat-related illnesses and deaths. The Earth Policy Institute compiled statistics from the affected countries and concluded that over 52,000 people died. Most of these deaths were not attributed to what epidemiologists call "short-term mortality displacement." These were not people whose health was already seriously compromised and expected to die soon who succumbed sooner because of heat-imposed stress. These were true excess deaths. Persons over 65 years of age, infants, and people with chronic medical conditions are at the greatest risk for the development of heat-related illnesses.

The European heat wave was not an isolated incident [8]. In the interval between 2001 and 2007, ten countries reported an increase in the number of heat-related deaths. In addition to highly developed nations, such as Canada and Switzerland, increases were reported in less well developed countries, such as Tajikistan. India experienced 18 heat waves between 1980 and 1998.

According to the data compiled by the National Oceanographic and Atmospheric Administration (NOAA) the summer of 2011 was the second hottest ever recorded. The mean daily temperature was a full three degrees higher than the average recorded during the twentieth century (available at: http://www.bitsofscience.org/us-summer-hottest -summer-record-2011-3132/). Texas, Louisiana, and New Mexico experienced the hottest summer ever recorded. Texas, southern California, and Arizona each experienced seventy or more days when the temperature was higher than 100 degrees Fahrenheit.

According to the Centers for Disease Control, 8,015 Americans died from heat-related illnesses between 1979 and 2003, more than died from hurricanes, tornadoes, lightning, floods, and earthquakes combined [13]. It is virtually certain that the morbidity associated with heat will increase as the planet gets warmer.

Heat-related illnesses develop when the intrinsic body mechanisms that control body temperature are unable to deal with heat from the environment or heat generated by exercise. This failure of the body's ability to cool itself leads to an uncontrolled elevation of body temperature. Sweating is a first-line of defense against heat. When sweat evaporates, the body is cooled. If body water stores drop too far, sweating may cease, and heat-related symptoms may develop.

The Centers for Disease Control has published the following guidelines concerning heat-related illnesses [13]. Heat cramps are the mildest of the heat-related illnesses. Cramps may also occur in patients with heat exhaustion, a more serious medical condition. Again, excessive loss of body fluids with inadequate replacement is an important cause. Symptoms of heat exhaustion include profuse sweating, cramps, fatigue, weakness, and pallor, which may be accompanied by dizziness, headache, nausea, and fainting. The skin may be cool and moist to the touch as blood vessels in the skin constrict. Heat stroke, a true medical emergency, is the most severe form of heat illness. Untreated patients may die or, if they recover, may be left with permanent disabilities. These patients are no longer able to regulate their body temperature. A temperature of 103 degrees Fahrenheit is needed to make the diagnosis, but temperatures as high as 106 degrees may be encountered. The skin is dry, red, and hot, due to an absence of sweat. The pulse is typically strong and rapid. Nausea and confusion are common, and some patients may become unconscious. It is imperative to cool these patients quickly by getting them out of the direct sun and into the shade or, preferably, into an air-conditioned environment. Unless patients are completely alert, attempts to give fluids by mouth may cause choking and create a risk of aspiration and secondary pneumonia.

Effects on Infectious Diseases

Infectious diseases, particularly if one includes waterborne and diarrheal illnesses, affect millions of people each year. According to the Intergovernmental Panel on Climate Change Fourth Assessment, there is a "high or very high [level of] confidence" that global warming will change the range of certain infectious disease vectors and affect malaria, and a "medium [level of] confidence that global warming will increase the burden of diarrheal illnesses." By expanding the range of insects that are carriers of diseases or disease vectors, the size of the population exposed to vector-borne diseases increases.

Dengue

Although the Intergovernmental Panel on Climate Change health panel concluded that there was a "low level" of confidence that global warming would increase the number of people who were at risk for developing dengue, other data suggest that there are still very good reasons to be concerned about this disease [8].

Dengue is one of the most prevalent, if not the most prevalent, vector-borne viral illness in the world [14]. It affects hundreds of millions each year and is transmitted primarily by the mosquito *Aedes aegypti*, an insect that is particularly well adapted to urban environments. Since there are no vaccines or drugs that are effective to prevent or treat the disease, public health measures must focus on control of the mosquito vector.

One and a half billion people live in an area where there is at least a 50% probability of contracting dengue [14]. This number is based on a model of geographical boundaries and limits for dengue transmission. Based on the best available data as many as 5 to 6 billion people will be at risk for developing dengue by 2085. This increase is based on increases in the amount of water vapor in the air that are expected to occur due to global warming. Atmospheric water vapor is highly correlated with dengue risk. If global warming could be averted, the population at risk would be lower, at around 3.5 billion individuals.

Dengue is also known as breakbone fever, a name that conveys a vivid image of the severe joint pain that may develop in individuals with severe forms of the illness. Dengue is caused by any one of four different but distinct forms of a *flavivirus* [15]. These viruses are in the same genus as the virus that causes West Nile Fever and Yellow Fever. Dengue is classified as being either uncomplicated or severe. Symptoms typically begin 4 to 7 days after infection with the sudden development of fever and headache that is frequently the most severe behind the eyes. Many infected patients recover spontaneously after a mild, self-limited illness that may not be diagnosed as dengue. Some develop very severe joint pain. Hence the name breakbone fever. Small hemorrhages that are between 1 and 2 mm in size may develop in the skin. Spontaneous recovery is the rule in uncomplicated dengue, but many patients complain of severe fatigue. The number of platelets in the blood is commonly reduced. Leakage of plasma out of blood vessels, bleeding, and impairment of organs, such as the liver, characterize severe dengue. The development of severe dengue or dengue hemorrhagic fever is heralded by the development of abdominal pain, enlargement of the liver, increases in the apparent number of red blood cells per unit volume (hematocrit), and a fall in the blood platelet count. Ironically, those who have been infected at an earlier time by one of the other forms of the dengue *flavivirus* are more likely to develop the severe form of the disease. This is thought to be the result of a phenomenon known as antibody-mediated immune enhancement.

Malaria

Malaria is one of the most serious of all diseases causing between 2 to 3 million deaths per year [16]. Children are particularly susceptible. The prevalence of malaria may be as high as 70% to 80% among individuals who live in areas where the disease is highly endemic.

The most recent Intergovernmental Panel on Climate Change report concludes with a "very high confidence" that climate change will "have mixed effects on malaria." In some places the range will expand and in others it will contract. Some of the factors that are responsible for this are: the complexity of the disease, nonclimatic factors such as socioeconomic development, drug-resistance and immunity, and the insufficiency of detailed observations of climate and malaria prevalence. The largest changes are the most likely to occur at the boundaries of locations where the disease is endemic rather than in the places where it is already rampant and can't get much worse.

Malaria is a protozoal disease caused by various species of *Plasmodium* carried by mosquitoes. The most severe form of malaria is caused by *P. falciparum*. Other forms are caused by infection by *P. vivax*, *P. ovale*, and *P. malariae*.

Malaria may have many manifestations [16]. The recurrence of fevers and chills in a regular cyclic fashion is characteristic. The duration of the cycle depends on the species of the protozoan. Coma is the most fearsome manifestation of malaria; this may be due to blockage of blood vessels by the protozoan or secondarily, by low levels of glucose in the blood. When malaria affects the kidneys, blackwater fever may develop, a condition characterized by a very dark urine caused by rupture of red blood cells and the excretion of hemoglobin. Pulmonary edema, an abnormal accumulation of fluid in the lungs that may be manifested by pink frothing at the mouth, may develop. Diarrhea may occur as a result of protozoal infections of the small blood vessels of the intestine.

At present, efforts to produce an effective vaccine have been unsuccessful. Control of the disease is usually the result of minimizing the number of mosquitoes that carry the protozoa by draining standing water, placing insecticide-impregnated nets over beds, and using mosquito repellents, such as DEET.

Diarrheal and Waterborne Illnesses

The United Nations Millennium Development Goals call for a two-thirds reduction in the mortality among children between the years of 1990 and 2015 [17]. Diarrheal diseases cause more deaths among children than malaria [17]. UNICEF estimates that diarrhea causes one of every five deaths in this age group, making it the second leading cause of deaths due to infectious diseases, claiming the lives of 1.5 million children each year [18].

The incidence of diarrheal diseases is highly dependent on climate and temperature. This is shown clearly by data acquired over a six-year period in a diarrheal disease clinic in Lima, Peru [19]. The authors reported a 4% increase in admissions for an increase of one degree centigrade during the hottest months and a 12% increase during the coldest months. The overall effect was an 8% increase per degree centigrade. This effect is likely to be multifactorial as temperature and humidity affect the rate of replication of pathogenic protozoal and bacterial organisms and the survival of environmental enteroviruses (a large family of viruses characterized by single strands of ribonucleic acid in their genetic sequence) [9].

Cholera is an excellent example of a diarrheal disease that can be linked to the effects of climate change [20]. Climate change induced flooding and disruptions in the social infrastructure, similar to those seen in Haiti in the wake of the combined effects of the earthquake and hurricanes, often set the stage for the emergence of cholera. Cholera is an acute diarrheal disease caused by an infection of the small intestine by the bacterium *Vibrio cholerae*. The infection is usually caused by consuming food or water that has been contaminated by the feces of an infected person. Direct person-to-person transmission is less common. Cholera affects persons of all ages during epidemics. In endemic areas it is often restricted to those over two years of age. Cholera is characterized by the abrupt onset of severe diarrhea, vomiting, cramps, and a marked decrease in the rate of urine production, secondary to dehydration. Patients do not usually have an elevated temperature. The stools may be characterized by the classic term "rice water." The disease may progress very rapidly with death occurring mere hours after the onset of symptoms. Untreated cholera has a death rate of 50% or more. Fortunately, less than 1% die if they receive adequate timely treatment. Oral rehydration therapy is usually sufficient to treat most patients. Intravenous fluid therapy and antibiotics may be required for those who are affected severely.

An epidemic of cholera developed in earthquake-ravaged Haiti. It is thought that the infection was introduced by aid workers. Because of the extreme poverty and the lack of an effective Haitian infrastructure, hundreds were affected. Rains and flooding caused by Hurricane Thomas exacerbated the problem. This concatenation of disasters is likely to be replicated in areas where cholera is endemic. Severe weather events caused by global warming are likely to create the conditions that favor the development of cholera epidemics.

Other Diseases

Other viral diseases may become more common as the result of global warming. In addition to carrying dengue, mosquitoes act as vectors for several forms of encephalitis including St. Louis and West Nile [21]. In contrast to other diseases associated with an increase in rainfall and floods, these two infections are associated with droughts. Hantavirus infections, usually transmitted via rodent urine, particularly that from deer mice, may be linked to El Niño related heavy rainfall in the southwestern states of the United States.

Leishmaniasis, a protozoal disease caused by many organisms in the genus *Leishmania*, may become more common as a consequence of global warming. It is transmitted via the bite of sand flies. This disease has many manifestations, most notably skin ulcers that erupt weeks or months after the bite. Months or even years later fever, anemia, or damage to the spleen may occur. Those affected by HIV/AIDS are likely to be much more susceptible. Since HIV/AIDS affects many individuals in various parts of the world, changes in the habitats for the animal vectors of the disease could lead to secondary changes in disease prevalence.

Leptospirosis is a relatively uncommon disease. The 1995 Nicaraguan epidemic of leptospirosis, a bacterial disease caused by spirochetes of the genus *Leptospira*, was highly linked to wading in flood waters contaminated by animal urine containing the bacteria. The bacteria gain entrance to the body via breaks in the skin, mucous membranes, or the eyes.

Effects on Agriculture and the Food Supply

Adequate nutrition is an absolute requirement for good health. According to the World Hunger Education Service, there were 925 million hungry people in the world in 2010: 578 million of these live in Asia and the Pacific region, and 239 million live in Sub-Saharan Africa [22]. It is virtually certain that this number will increase as the result of global warming.

The very large number of people who are already starving, uncertainties in the estimates of the rate of increase in the world's population, and the importance of growing enough food to feed everyone make it extremely difficult to predict future agricultural yields. This is perhaps one of the most daunting of all modeling tasks facing scientists who seek to understand the implications of climate change.

In order to set the stage for predicting the effects of climate change on crop yields, a group of scientists recently investigated the effects of temperature and precipitation on the worldwide yields of wheat, maize, soybeans, and rice [23]. The authors noted that these four crops account for about 75% of all of the calories that people consume either directly or indirectly (e.g., by feeding animals raised for food). For their study they utilized publicly available data concerning crop yields for virtually all countries in the years 1980 to 2008—almost three decades. They combined these data with records of temperature and precipitation to model crop yields that would be expected in the absence of climate change. There were significant differences between the observed and expected global yields for maize, which was reduced by 3.8%, and wheat, which was reduced by 5.5%. There was little effect on the yields of rice and soybeans. This does not bode well for the future. The effects of climate change more than offset increases in yields that were expected to occur as the result of other factors, including increases in the concentration of CO_2 in the atmosphere, hardier varieties, and fertilization.

The 2004 World Health Organization's attempt to quantify health risks associated with climate change relied heavily on the work of Parry et al. [9,24]. These investigators have developed increasingly sophisticated models that have been adapted for use in predicting crop yield at various locations and the effects on the number of people at risk for the development of malnutrition.

These models include baseline data on climate and crop yields from much of the developed and developing world, with the exclusion of China. These data were used to estimate parameters such as the impact of improved crop genetics, temperature, and other factors, on yields.

These parameters, based on the reference baseline observations made between 1961 and 1990 were combined to estimate the distribution of yields for three critical crops: corn, wheat, and soybeans. Insufficient data from southeast Asia and China precluded modeling the impact on rice production.

Although for the purpose of this book, complete details of the modeling strategy are somewhat irrelevant to the result, a few details may help one understand the level of sophistication of the models and their limitations. For example, when plants are grown under experimental conditions, where the concentration of CO_2 can be controlled, it is possible to detect and quantify an increase in the efficiency of photosynthesis, which improves crop yields. In addition the efficiency with which most plants use water also increases. This is due to the fact that the small openings in leaves, known as stomata, close partially as the atmospheric CO_2 concentration rises. CO_2-enhanced growth and lower water loss via stomata are both beneficial. Other factors act to decrease crop yields, particularly those grown near their temperature limits. Higher temperatures reduce the length of the growing period for some crops, decreasing the length of time during which the nutritional value of grains develops. This, in turn, reduces the nutritional value of the crop. Projected reductions in the availability of water due to changes in precipitation, water loss from soils, and other factors will also have adverse effects on crop yields. Finally, yields from plants that require a period of low temperatures during the winter, such as winter wheat, will suffer.

If climate change were to have no effect on crop yields, world cereal production would be expected to rise from the approximately 2.2 million tons produced in 1990 to about twice that amount by 2080 [24]. However, a shortfall of about 4% is predicted by the model. Due to the shortage, food prices are predicted to rise by over 40%. These combined effects will cause substantial increases in the prevalence of malnutrition. Since malnutrition makes people susceptible to a variety of diseases, disease-related mortality and morbidity will also increase. These effects will cause increasing disparities between nations located at higher latitudes, where there may actually be increases in yields, and nations at lower latitudes, namely the developing nations. Climate-induced reductions in crop yields may be as high as 16.5% in these nations. It is these people who will suffer the most. They already bear the greatest burdens associated with hunger and malnutrition.

These are likely to be optimistic estimates. For example, the models assume adequate control of pests, such as weeds and insects, availability of enough fertilizers, and the absence of significant droughts and floods. The models also fail to account for the price increases that have been caused by diverting corn from the food supply to ethanol production for use in motor vehicle fuels. As of the time of this writing, unrest in the Middle East has had an unpredictable effect on oil production. Further disruptions have the potential to confound costs associated with raising crops in highly industrialized settings.

As one might imagine, it is very difficult to predict the effects of climate change on agriculture. However, based on the best current complex models of climate and agriculture, it is likely that 80 million more individuals will join the ranks of the malnourished by 2080 [24].

Coastal Flooding

The combination of migration toward coastal areas and rising sea level creates the potential for widespread flooding of inhabited areas. Coastal areas are already experiencing many of these effects [25]. At present approximately 120 million people are potentially vulnerable to storms, particularly tropical cyclones and hurricanes. It is thought that these events caused 250,000 deaths among those at risk between 1980 and 2000 [25]. As migration toward these areas continues, sea levels rise, and extreme weather events increase in frequency, it is predictable that more deaths and population displacements will occur as the result of global warming. Figure 12.3 illustrates many of these interdependencies.

The health effects of flooding go beyond the need to move large numbers of people and deaths due to drowning. Climate refugees are more likely to move to locations where the infrastructure is inadequate to meet basic needs for sanitation and the delivery of safe drinking water. Many river deltas are important agricultural areas or sites where fishing and other forms of aquaculture provide food.

Ericson and his associates studied the effects of increases in sea level on 40 of the largest deltas in the world [26]. They report that 300 million people live in these deltas. Using current measurements of the rate at which the effective levels of the oceans are rising, they predict that 8.7 million people who live on 28,000 km^2 of these deltas will suffer displacement by 2050. This flooding will be due to the combined effects of inundation by seawater and predicted increases in the erosion of these deltas. These estimates do not include the effects that are likely to occur as the result of storms and storm surges.

The Nile, Ganges-Brahmaputra, and Mekong Deltas are at extreme risk, defined as having a population of more than one million that is subject to displacement [26]. The Godavari Delta in India, the Mississippi Delta in the United States, and the Yangtze or Chang Jian Delta in China are classified as being at high risk, meaning that between 50,000 and one million people will be subject to displacement due to flooding [26].

The Intergovernmental Panel on Climate Change report concludes that the optimum way to deal with the inevitable rise in sea level will require a combination of mitigation and adaptation. Mitigation, in terms of limiting greenhouse gas emissions, is required to minimize temperature and sea level increases. Adaptation is required to minimize the impact of the inevitable increases in sea level.

Although both adaptation and mitigation will be expensive, the prediction is that the benefits will exceed the cost. The Intergovernmental Panel on Climate Change Second Working Group report on changes in coastal systems includes several cost–benefit estimates. In order to protect nations that are a part of the Organization for Economic Cooperation and Development against a 1-meter rise in seal level, mitigation costs are expected to reach $136 billion. This would prevent the displacement of around 220,000 people. Benefits are expected to exceed costs by a factor of 10 if the inhabitants the Pearl River Delta in China are to be protected from the effects of 0.65-meter rise in sea level. For the United States, the benefits may be as high as five times the cost to guard against a one meter rise in sea level.

The dikes in the Netherlands and the Thames Barrier in London are examples of existing flood protection. Thus, under some circumstances, there is a demonstrable political will to protect citizens from the effects of rising sea level. However, it remains to be seen whether it is possible to generate the will to protect coastal areas from the threats posed by rising sea level due to global warming. For too many, this threat is too easily dismissed.

Global Warming and Air Quality

Global warming is predicted to have important effects on the concentration of ground-level ozone. Other aspects of air quality are more difficult to estimate.

As described in detail in chapter 3, ground-level ozone is formed by a chemical reaction between oxides of nitrogen and volatile organic compounds. Sunlight and heat provide the energy to drive this reaction. From temperature and air quality monitoring data, ambient temperatures and atmospheric ozone concentrations are highly correlated: an increase in temperature is typically associated with an increase in the ozone concentration. This is most prominent in summer months. Thus the increase in the number of heat waves and general increases in the temperatures that are expected across the Northern Hemisphere are almost certain to lead to more days when ozone levels pose a threat to health.

As the earth warms, there will be changes in the patterns and velocity of the wind. This, in turn, will affect the dispersion of various air pollutants. Particulate matter will be one of the pollutants affected. The behavior of particulates is the subject of substantial uncertainty. On one hand, the increase in the amount of heat energy will tend to favor some of the chemical reactions that form secondary particulates. This may increase their concentration. On the other hand, the formation of the aerosols that contain particulates blocks solar energy and tend to decrease solar warming. The balance between these opposing forces makes it difficult to include particulates in global warming models.

Other factors that contribute to air quality may also be affected. Forest fires associated with drought may become more numerous and inject soot into the air along with other pollutants, including mercury. Winds blowing across deserts will pick up and disperse more dust. The magnitude of these factors is likely to vary substantially in different parts of the world making their effects difficult to predict.

Science, Perception, and Global Warming

Science is a hypothesis-driven discipline that is self-correcting by its very nature. An idea generates a hypothesis that is then subjected to testing by experimentation. This either strengthens or weakens the hypothesis that is revised and is again subjected to experimentation. Indeed there are those who rightly contend that if an idea or hypothesis can't be subjected to experimentation, it falls outside of the realm of science. The hypothesis that the earth is warming slowly due to human activity is an example of a hypothesis that not only *can* be tested but *has* been tested time and time again, refined, and retested. At present, there is virtually no credible evidence that the global warming hypothesis is not true. However, it is true that, as with all of science, there are limits on the confidence with which specific conclusions may be drawn, for example, *exactly* how much warming can we expect and *how much* will sea level rise by a given date? Even so, an extraordinarily high level of confidence can be placed on the predictions that mean temperatures and sea level will increase if steps are not taken soon to halt these seemingly inexorable events.

There is an excellent example of the self-correcting nature of science in a chapter on the history of global warming [27]. Several articles written in the 1970s by respected scientists predicted that the cooling effects of particulate emissions and the formation of atmospheric aerosols would offset the warming effects of increasing levels of CO_2. This hypothesis was subjected to experimental testing. The results led to the rejection of the aerosol-offset hypothesis when it was shown that aerosols remained in the atmosphere for a short time, relative to the much longer residence time for CO_2. It was therefore deemed to be inappropriate to include the effects of aerosols in long-range climate projections. This conclusion has been borne out by subsequent observations that have shown that global temperatures have continued to rise, even as the direct emission of particulates increased and the amount of particles formed secondarily increased.

This is but one small example of how science and global warming have meshed. As scientists have studied the myriad details, each of which contributes to our understanding of global warming, predictions about the future have become more and more refined. With this increasing refinement there is a corresponding certainty that the threat to humanity is real and grave. It remains to be seen whether we are able to muster the will to act and take the steps that many scientists, economists, and others say are needed to mitigate the impending disaster. This may be a defining moment for us as a human species.

References

1. IPCC Core Writing Team. IPCC, 2007: Climate Change 2007: Synthesis Report. Contribution of Working Groups I, II and III to the Fourth Assessment Report of the Intergovernmental Panel on Climate Change. Geneva: IPCC, 2007.

2. Forster P, Ramaswamy V, Artaxo P, et al. Changes in atmospheric constituents and in radiative forcing. In: Solomon S, Qin D, Manning M, Chen Z, Marquis M, Averyt KB, et al., eds., Climate Change 2007: The Physical Science Basis. Contribution of Working Group I to the Fourth Assessment Report of the Intergovernmental Panel on Climate Change. Cambridge: Cambridge University Press, 2007:130–234.

3. Oak Ridge National Laboratory Carbon Dioxide Information Analysis Center. United Nations Millennium Development Goals Indicators: Carbon Dioxide Emissions. Oak Ridge TN, 2010. Available at: http://cdiac.ornl.gov.

4. Keeling RF, Shertz SR. Seasonal and Interannual Variations in atmospheric oxygen and implications for the global carbon cycle. Nature 1992;358:723–7.

5. Gannes LZ, Martinez del RC, Koch P. Natural abundance variations in stable isotopes and their potential uses in animal physiological ecology. Comp Biochem Physiol A Mol Integr Physiol 1998;119(3):725–37.

6. Prentice IC, Farquhar Gd, Fasham MJR, et al. The carbon cycle and atmospheric carbon dioxide. In: Houghton JT, Ding Y, Griggs DJ, Noguer M, van der Linden JP, Dai X et al., eds., Cambridge: Cambridge University Press, 2011: 184–238.

7. US Energy Information Administration. Voluntary Reporting of Greenhouse Gases Program: Fuel Emission Coefficients. Washington DC: Government Printing Office, 2011.

8. Confalonieri U, Menne B, Akhtar R, et al. Human health. In: Parry ML, Canzizni OF, Palutikof JP, van der Linden JP, Hanson CE, eds., Climate Change 2007: Impacts, Adaptation and Vulnerability. Contribution of Working Group II to the Fourth Assessmemt Report of the Intergovernmental Panel on Climate Change. Cambridge: Cambridge University Press, 2007:391–431.

9. McMichael AJ, Campbell-Lendrum D, Kovats S, et al. Global climate change. In: Ezzati M, Lopez AD, Rodgers A, Murray CJL, eds., Comparative Quantification of Health Risks: Global and Regional Burden of Disease Atrributable to Selected Major Risk Factors, vols. 1, 2. Geneva: WHO, 2004:1543–1649.

10. Oxford English Dictionary, 2nd ed. Oxford: Oxford University Press, 2002.

11. Frich P, Alexander LV, Della-Marta P, et al. Observed coherent changes in climatic extremes during the second half of the twentieth century. Clim Res 2002;19:193–212.

12. Larsen J. Setting the Record Straight: More Than 52,000 Europeans Died from Heat in Summer 2001. Washington DC: Earth Policy Institute, 2006.

13. Centers for Disease Control and Prevention. Extreme Heat: A Prevention Guide to Promote Your Personal Health and Safety. Atlanta: CDC, 2011.

14. Hales S, de WN, Maindonald J, Woodward A. Potential effect of population and climate changes on global distribution of dengue fever: an empirical model. Lancet 2002;360(9336):830–4.

15. Whitehorn J, Farrar J. Dengue. Br Med Bull 2010;95:161–73.

16. Krogstad DJ. Malaria. In: Goldman L, Ausiello F, eds., Cecil Medicine, 23rd ed. Philadelphia: Saunders, Elsevier, 2008:2375–80.

17. Department of Economic and Social Affairs of the United Nations Secretariat. The Millennium Development Goals Report: 2010. New York: United Nations, 2010.

18. UNICEF. Diarrhea: Why Children Are Still Dying and What Can Be Done. Geneva: WHO, 2009.

19. Checkley W, Epstein LD, Gilman RH, et al. Effect of El Niño and ambient temperature on hospital admissions for diarrheal diseases in Peruvian children. Lancet 2000;355(9202):442–50.

20. Gotuzzo E, Seas C. Cholera. In: Goldman L, Ausiello F, eds., Cecil Medicine, 23rd ed. Philadelphia: Saunders, Elsevier, 2008:2227–30.

21. Haines A, Patz JA. Health effects of climate change. JAMA 2004;291(1): 99–103.

22. World Hunger Education Service. 2011 World Hunger and Poverty Facts and Statistics. Washington DC: Government Printing Office, 2011.

23. Lobell DB, Schlenker W, Costa-Roberts J. Climate trends and global crop production since 1980. e-Published ahead of print. Science 2011.

24. Parry M, Rosenzweig C, Iglesias A, Fischer G, Livermore M. Climate change and world food security: a new assessment. Global Environ Change 1999; 9:S51–67.

25. Nicholls RJ, Wong PP, Burkett JO, et al. Coastal systems and low-lying areas. In: Parry ML, Canzizni OF, Palutikof JP, van der Linden JP, Hanson CE, eds., Climate Change 2007: Impacts, Adaptation and Vulnerability. Contribution of Working Group II to the Fourth Assessment Report of the Intergovernmental Panel on Climate Change. Cambridge: Cambridge University Press, 2011: 315–56.

26. Ericson JP, Vörösmarty CJ, Dingman SL, Ward LG, Meybeck M. Effective sea-level rise and deltas: causes of change and human dimension implications. Global Planet Change 2006;50(1–2):63–82.

27. Le Treuet H, Somerville R, Cubasch U, et al. Historical overview of climate change. In: Solomon S, Qin D, Manning M, Chen Z, Marquis M, Averyt KB, et al., eds., Climate Change 2007: The Physical Science Basis. Contribution of Working Group I to the Fourth Assessment Report of the Intergovernmental Panel on Climate Change. Cambridge: Cambridge University Press, 2007:95–127.

13

Energy and Health Care Economics

A penny saved is a penny earned.
—Benjamin Franklin

We are caught in the winds of a perfect storm. Health care costs are rising faster than any other major segment of the economy, the national debt threatens to become unmanageable, the gap between the rich and the poor is getting wider, and the number of people living below the poverty line in the United States is higher than ever. All of this is occurring at a time when we are struggling to emerge from the worst financial crisis since the Great Depression. Congressional pressures to slash funds from discretionary portions of governmental budgets have become irresistible. This, in turn, threatens the vision that Barack Obama put forth in his 2011 State of the Union speech where he called on the US public to "out-innovate, out-educate, and out-build the rest of the world." Oddly enough, burning coal, clean air, and health could and should play central roles in the debates that have erupted as we struggle to deal with the consequences of this financial maelstrom.

Estimating Health Impacts

To assess the effect of a pollutant on a health endpoint, such as mortality or hospital admission for a given disease, knowledge of a number of basic facts is needed. The EPA outlined some of these requirements in their most recent cost–benefit analysis [1]. First, one must define and know something about the population that is potentially affected. This might be an entire nation, or it might be a smaller segment of the population, such as those living downwind of a given power plant in a census tract or area within a zip code. These statistics are maintained by the US Census Bureau, and similar agencies. Second, it is necessary to know something about the baseline incidence of the health effect. The Centers for Disease Control and Medicare statistics are sources for this information. Examples of these statistics were presented in chapters 8, 9, and 10. Third, a dose–response or concentration–response function must be obtained from the existing epidemiological literature. This variable will describe the expected change in the endpoint, such as the number of admissions to the hospital for a stroke, for a given increment/decrement in the concentration of the pollutant. The fourth element is an estimate of the change in the concentration of the pollutant of interest. These data are collected at monitoring sites established by the EPA and other agencies. Finally it is necessary to monetize the value of each case avoided. The health economics literature can be used for this purpose, along with other sources, such as Medicare payments to hospitals that are based on a given diagnosis.

Given all of the factors involved, estimating the health impacts attributable to a source of an environmental pollutant is difficult. Substantial efforts have been devoted to the development of various computer modeling strategies that are used to perform this task. These models have become more accurate, sophisticated, and valid as time has passed, due to a number of advances. First, it is possible to test the model by beginning with a given date and comparing the predicted effect, such as the change in the concentration of a specific pollutant at a given time, with an actual measurement of the concentration at that time. Model parameters can then be adjusted so that the predicted concentration and measured actual concentration converge. This strategy is employed routinely by almost every model developer, regardless of the parameter to be modeled. Second, the data that are fed into the model are becoming more refined. This is due in part to the fact that the EPA and other agencies have established a nationwide network of monitoring stations that are designed to collect pollutant concentrations at given times at specific locations. The location and concentration data from these monitoring sites are available from the EPA Office of Air and Radiation [2]. Third, there is much more specific information on concentrations and health responses available in the epidemiological literature. As statistical methods have improved, it has become possible to link even brief changes in the concentration of a pollutant, such as fine particles, with a change in a health endpoint, such as hospitalization for an acute myocardial infarction, as discussed in chapter 9.

Specific Cost–Benefit Analyses

The Clean Air Act and Other EPA Regulations

As a part of the 1990 amendments to the Clean Air Act, Congress required the EPA to conduct "periodic, scientifically reviewed studies to assess the benefits and the costs of the Clean Air Act" [3]. In other words, Congress wanted to know whether the Act "was worth it." The initial report in what is now a series was released in October, 1997. The evaluation provided a detailed retrospective analysis of costs and benefits from the years 1970 to 1990 and showed that the overwhelming benefits obtained from compliance with the Act far outweighed the costs of implementation.

In the analysis the EPA used dose–response data from the scientific literature available at the time to estimate the effects of a given increment or decrement in the concentration of a given pollutant in the air. They combined these data with values associated with a given health outcome from economics data, reported in terms of cost per case, episode, symptom-day, and so forth. These data were combined with actual measurements of air quality and estimates of air quality if the controls had not been in place. Using the modeling strategies available at the time, the EPA estimated that the Act resulted in a 40% reduction in sulfur dioxide, a 30% reduction in oxides of nitrogen, a 50% reduction in carbon monoxide, and a 45% reduction in total suspended particles. The reductions in fine and larger particles were thought to be on the order of 45%, based on measurements of total suspended particles. Specific monitoring for particulate matter fractions had not been instituted at that time, so an extrapolation from total suspended particles was required.

Some of the specifics from the analysis are shown in table 13.1. The improvements in air quality were thought to be primarily due to reductions in particulate matter and ozone. In this retrospective analysis, the modeling predicted an annual reduction of 184,000 premature deaths, 674,000 cases of chronic bronchitis, over 22 million lost days at work, and other outcomes shown in the table.

Table 13.1
Health effects avoided by the Clean Air Act, 1970 to 1990

Effect	Pollutant(s)	Annual effect avoided	Effect value per case
Premature mortality	PM	184,000 cases	$4,800,000
Chronic bronchitis	PM	674,000	$260,000
Ischemic heart disease hospitalization	PM	19,000	$10,300
Congestive heart failure	PM and CO	39,000	$8,300
COPD and pneumonia hospitalization	PM and ozone	62,000	COPD: $$8,100; Pneumonia: $7,900
Acute asthma attack	PM and ozone	850,000	$32
Days lost work	PM	22,600,000	$83 per day

Note: PM = particulate matter, CO = carbon monoxide, COPD = chronic obstructive pulmonary disease.

The EPA concluded that the total monetized health benefits from the Act during the twenty-year period ranged between *$5.6 and $49.4 trillion.* The central estimate for benefits was $22.2 trillion. During that period, the costs to comply with the act were estimated to be approximately $0.5 trillion. Thus the net direct benefits were between $5.1 and $48.9 trillion, with a central estimate of $21.7 trillion. The benefit–cost ratios were 43.4:1 for the central estimate and 11:1 and 97.8:1 for the extreme estimates. Who among us has an investment that has performed this well?

The second prospective EPA cost–benefit analysis was released in March, 2011 [1]. The results of this study reflect the vast improvements in our understanding of the effects of particulate matter on the risk of premature death. These improvements are a direct result of the publication of large epidemiological studies and emphasize the importance of continuing research in this area. In addition the second prospective report adds several endpoints such as changes in visibility due to improved air quality, a better understanding of the relationship between peaks in fine particle concentrations and acute myocardial infarcts, and better mathematical modeling of air quality. A summary of these projections, based

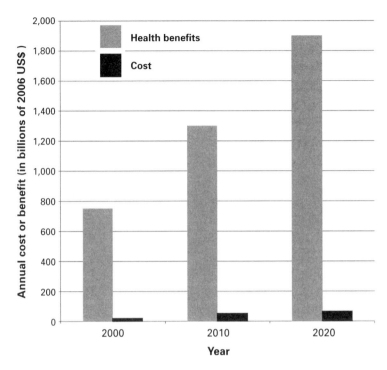

Figure 13.1
Annual benefits and costs of the Clean Air Act from 1990 to 2020 [1].

on central estimates of both costs and benefits, is shown in figure 13.1. The benefit–cost ratio for the central estimates ranges from a low of 25:1 for 2010 to a high of 39:1 for 2000. Benefit–cost ratios, based on the high estimates, range from 72:1 for 2010 to 115:1 for 2000. Any way one may choose to interpret these data, benefits consistently outweigh costs by very large margins.

The projected benefits are attributed primarily to reductions in the concentrations of ground-level ozone and fine particles. The EPA estimates that in 2020 the Clean Air Act amendments will result in a 17% reduction in the direct emissions of small particles, or a reduction from 6,368 to 5,297 tons. While reductions in the emission of these primary particulates are a positive development, the most significant reductions in the total concentration of fine particles are attributed to reducing sulfur dioxide emissions. Atmospheric sulfur dioxide combines rapidly with other compounds in the atmospheric aerosol to form secondary

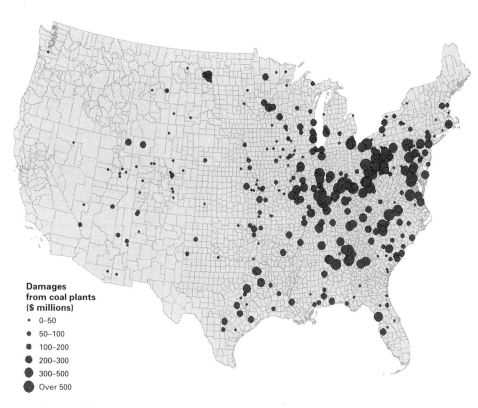

**Damages
from coal plants
($ millions)**

- 0–50
- 50–100
- 100–200
- 200–300
- 300–500
- Over 500

Figure 13.2
Health care costs, or damages, associated with 406 coal-fired power plants [6].
Used with permission of the National Academy of Sciences.

particulates, as described in chapter 3. Thus sulfur dioxide emissions are a significant source of the particulates that the study identifies as the most important cause of adverse health effects. In the report the Agency predicts that sulfur dioxide emissions in 2020 with the amendments in place will be just over 8,200 tons per year, as opposed to a predicted 27,900 tons without the amendments. Coal-fired power plants are by far the most important source of this pollutant [4]. About 75% of the reductions in sulfur dioxide releases will come from curtailing this source. Year 2020 emissions of nitrogen oxides, with the amendments, are predicted to be about 10,000 tons as opposed to about 31,700 tons without the amendments. As described in detail in chapter 3, ground-level ozone is formed by the chemical reaction of these oxides of nitrogen and volatile organic compounds in the presence of sunlight. The National Ambient Air Quality website, maintained by the EPA, lists electrical generating units as the third leading source of oxides of nitrogen [4]. Utilities trailed on-road vehicles and non-road equipment that emitted 6.5 and 4.2 million tons of nitrogen oxides, respectively.

Reductions in premature deaths are the most important source of the monetized benefits associated with the Clean Air Act amendments. By the year 2020 the scenario predicted by the amended Act avoids 230,000 premature deaths among adults age 30 and above each year. The model also predicts avoiding the deaths of 280 infants each year. The monetary value of these two causes was set at $1.7 trillion for adults and $2.5 billion for infants. Reductions in the number of cases of bronchitis, asthma, myocardial infarction, and other health effects contribute to the predicted $2 trillion dollars in annual benefits by the end of this decade. A tabulation of the projected benefits, in terms of instances or cases avoided, is shown in table 13.4.

In the second prospective report, EPA scientists estimate that approximately $10 billion of the $65 billion in annual costs in 2020 will be borne by electrical utilities [1]. Based on what is known about the releases of oxides of sulfur and nitrogen, primary particulates, and the formation of secondary particulates, the benefit–cost ratio associated with a focus on this sector should yield benefit-to-cost ratios that are above the mean predicted by the model and shown in figure 13.1.

Other agencies of the federal government have examined the costs and benefits of EPA rules. In 2003 the Office of Management and Budget reported that the fiscal year 2001 costs associated with complying with unfunded EPA mandates and regulations were about $192 million yielding benefits of about $1.25 billion (in 2001 dollars) [5]. The estimates for the decade beginning October 1, 1992, included benefits ranging between $120.8 and $193.2 billion with cost estimates that ranged between $23.4 and $26.6 billion. The majority of these benefits and costs were attributable to rules promulgated under the authority of the Clean Air Act and its amendments.

Clearly, the analysis provided in the EPA second prospective study and by others justifies focusing efforts to improve air quality and health on coal-fired electrical generating units.

Hidden Costs of Burning Coal

Three important analyses of the hidden costs of burning coal have been published in the past several years. Even in the relatively short time span of the two years that separated the reports, predicted costs have risen dramatically.

In the Energy Policy Act of 2005 (Public Law 109-58) Congress required the Department of the Treasury to commission the National Academy of Sciences "to define and evaluate the health, environmental, security, and infrastructural costs and benefits associated with the production and consumption of energy that are not or may not be fully incorporated into the market price of such energy." The result of the study, titled *Hidden Costs of Energy: Unpriced Consequences of Energy Production and Use*, was published by the prestigious National Research Council (NRC) in 2009 [6]. Based on 2005 emissions and a value of a statistical life of $6 million, the NRC reported that the health damages attributable to 406 coal-fired electrical utilities were $62 billion per year (in 2007 dollars), or $156 million per plant, if divided equally. Figure 13.2 shows costs associated with each of the plants superimposed on a map of the United States. As shown in table 13.2, sulfur dioxide accounted for most of the damages. This table lists the average total

Table 13.2
Damages associated with coal-fired utilities

	National Research Council Study [6]			Levy et al. [7]
Pollutant	Average total cost (2007 US dollars)	Average cost per ton (2007 US dollars)	Average cost per kWh (2007 US cents)	Range, cost per ton
SO_2	$1.6 \times 10^8 \pm 1.9 \times 10^8$	$5,800 \pm 2,600$	3.8 ± 4.1	6,000–50,000
NO_x	$1.1 \times 10^7 \pm 1.1 \times 10^7$	$1,600 \pm 780$	0.34 ± 0.38	500–15,000
$PM_{2.5}$	$9.0 \times 10^6 \pm 1.3 \times 10^7$	$9,500 \pm 8,300$	0.30 ± 0.44	30,000–500,000
PM_{10}	$5.2 \times 10^5 \pm 6.9 \times 10^5$	460 ± 380	0.017 ± 0.023	Not calculated

Note: SO_2 = sulfur dioxide, NO_x = nitrogen oxides, $PM_{2.5}$ = particles with an aerodynamic diameter of 2.5 μm or less, PM_{10} = particles with an aerodynamic diameter of 10 μm or less.

damages, average cost per ton, and the average cost per kilowatt hour associated with four pollutants: sulfur dioxide, nitrogen oxides, and particles with diameters of 10 and 2.5 microns. The greatest damages were in the eastern part of the United States, particularly along the Ohio River, the middle Atlantic states, the western parts of the Carolinas and northern Mississippi, Alabama, and Georgia. The coal plants in these areas are the oldest and are the least likely to be equipped with modern pollution control devices. In addition many of these plants are located near population centers and expose large numbers of people to their pollutants.

As shown in table 13.3, the adverse health costs associated with these four pollutants produced by using natural gas as the energy source were several orders of magnitude lower than those associated with burning coal.

In the same year that the National Research Council released its findings, Levy et al. published an independent assessment of the costs attributable to burning coal [7]. This second study focused on the between-plant differences in health costs that were reported to vary between $0.02 and $1.57 per kilowatt hour of electricity. Emissions from many of the plants they studied have dropped substantially since 1999, due to the installation of pollution control equipment. Maps that depict the cost per ton for emissions of particulates, sulfur dioxide, and nitrogen oxides are a unique and interesting feature of the Levy report. Although the maps are, in general, somewhat similar to the data shown in figure 13.2, the costs

Table 13.3
External damages associated with coal and natural gas fueled boilers in electricity generating units [6]

	Fuel	
	Coal (406 plants)	Natural gas (498 plants)
Pollutant	Average total cost (2007 US dollars)	Average total cost (2007 US dollars)
SO_2	$1.6 \times 10^8 \pm 1.9 \times 10^8$	$6.40 \times 10^4 \pm 2.58 \times 10^5$
NO_x	$1.1 \times 10^7 \pm 1.1 \times 10^7$	$5.49 \times 10^5 \pm 1.25 \times 10^6$
$PM_{2.5}$	$9.0 \times 10^6 \pm 1.3 \times 10^7$	$8.31 \times 10^5 \pm 3.23 \times 10^6$
PM_{10}	$5.2 \times 10^5 \pm 6.9 \times 10^5$	$4.47 \times 10^4 \pm 1.75 \times 10^5$

Note: See table 13.2 for abbreviations

Table 13.4
Health effects avoided by ozone and $PM_{2.5}$ concentrations attributable to Clean Air Act amendments [1]

Health effect	Pollutant(s)	Year 2010	Year 2020
Adult mortality	PM	160,000	230,000
Infant mortality	PM	230	280
Ozone mortality	Ozone	4,300	7,100
Chronic bronchitis	PM	54,000	75,000
Acute bronchitis	PM	130,000	180,000
Acute myocardial infarction	PM	130,000	200,000
Asthma exacerbation	PM	1,700,000	2,400,000
Hospital admission	PM, ozone	86,000	135,000
Emergency room visits	PM, ozone	86,000	120,000
Restricted activity days	PM, ozone	84,000,000	110,000,000
Lost days in school	Ozone	3,200,000	5,400,000
Lost days at work	PM	13,000,000	17,000,000

Note: PM = particulate matter.

are more widely dispersed among the utilities. Not surprisingly, they found that a variable they called the intake fraction, "the fraction of a pollutant or its precursor that is inhaled by a member of the population" was the most important determinant of cost. Some cost data from this paper are included in table 13.2.

In the early part of 2011, a large group headed by Epstein published what is probably the most comprehensive analysis of the hidden costs of using coal [8]. They conclude that using coal costs the US economy around $345 billion each year (low and high estimates were $175 billion and $523 billion). In addition to the expected cost accounting associated with postcombustion emission estimates, they included all aspects of coal mining with a focus on mountain top removal, transport, and ash disposal. The impact on Appalachian ecosystems received special emphasis for two reasons: this region is impacted heavily by all aspects of coal use and the diversity of species is higher in this area than in most parts of the country. They also drew attention to the effects of nitrogen dioxide deposition in the water where this pollutant acts as a fertilizer that promotes harmful algal blooms and contributes to the formation of some 350 dead zones that threaten coastal fisheries.

Epstein et al. include a brief re-evaluation of the National Research Council Report [8]. They contend that if the Council had used other dose–response data, the $62 billion cost would have tripled to about $187 billion per year.

Do It Yourself Cost Analysis

The models used by the EPA and other organizations are very complex and not easy to understand by those who are not well versed in their use. However, it is possible to make an approximate, first-order estimate of costs attributable to a given power plant in the United States from data that are available to the general public. Emissions data are released by the EPA in the Emissions and Generation Resource Integrated Database, or eGRID [9]. A more user-friendly source that has less recent data on sulfur dioxide, nitrogen dioxide, and particulate emissions can be found on the website maintained by the EPA Office of Air and Radiation [2]. The estimate of the facility-specific damages can be made by summing the products of these online emissions data and cost per ton or cost per kWh damage data shown in table 13.2. This computation yields a mean estimate for any utility. It does not account for exposures to the public that are influenced greatly by the location of the facility. If the facility is located in or near a densely populated area, an upward adjustment of the damages would be necessary. One might guess at this by increasing the damage per ton figure by one or two standard deviations, using data shown in the table. Similarly utilities located in remote, relatively unpopulated sites would cause lower exposures because the population affected is small. This would require a downward adjustment of damages as reflected in the data of Levy et al. discussed above [7].

What's Not Here

The studies cited above all rely on very conservative predictions of costs and benefits. Each one has withstood the tests of rigorous review by advisory panels or journal editors. It is likely that significant upward projections of benefits will occur in future analyses. This was certainly the case when the 1970 to 1990 retrospective study was compared to the 1990 to 2020 prospective study. These adjustments were due, as discussed, to improved models, pollution monitoring data, and more recent epidemiological studies. It is likely that further improvements will occur in all of these areas.

Cost data are more likely to remain static than benefit data as long as pollution-control technology remains relatively fixed. However, drastic upward revisions in costs would be required if carbon capture and storage technologies were implemented on the scale needed to achieve significant control of carbon dioxide emissions.

As epidemiological methods have become more sophisticated and more is learned about disease mechanisms, the spectrum of diseases associated with particulate matter has increased. As discussed in chapter 11, there are reasons to suspect that type 2 diabetes and Alzheimer's disease, two common chronic diseases that are very expensive to manage, may be linked to particulates. If these links become certain enough to include these diseases in the list of pollution-related medical problems, benefit estimates are likely to soar.

Finally none of the studies that I have reviewed contain any estimate of the health costs that will inevitably be associated with global warming. Global warming is happening now and will continue to happen for the foreseeable future. The financial impacts of global warming will be large, and if history is a good teacher, there will be large benefits associated with controlling global warming. Costs associated with global warming are difficult to compute. However, Stern has estimated that prompt action to curtail greenhouse gas emissions and their effects will cost about 1% of the entire global gross domestic product (GDP) [10]. He estimates that a failure to act could lead to costs of 5% of the global GDP each year, "now and forever" with a worst case scenario that includes a wider range of potential impacts escalating the cost to 20% or more of the GDP!

Conclusions

Every study that has examined benefits and costs associated with curtailing sources of air pollution has shown that the benefits outweigh the costs by huge margins. In its second prospective report the EPA makes the following statement, ". . . the results suggest that it is extremely unlikely that costs of the 1990 Clean Air Act Amendment programs would exceed their benefits under any reasonable combination of alternative assumptions or methods which could be identified. Even if one were to adopt the extreme assumption that the fine particle and ozone pollution have no effect on premature mortality risk—or that such risk reductions occur but they have no value—the benefits of reduced, non-fatal health effects and improved visibility alone add up to $137 billion for the year 2020, an amount which is more than twice the estimated $65 billion cost to comply with all 1990 Clean Air Act Amendment requirements in that year" [1].

According to data cited by the Council of Economic Advisors in 2009, 18% of Americans under the age of 65 are Medicare or Medicaid beneficiaries or receive health care from the military, 59% have employer-sponsored health insurance, and 16% are not insured [11]. For Americans above the age of 65, the Centers for Disease Control report that 94% are enrolled in Medicare [12]. In other words, large portions of all health care costs are borne by federal, state, and municipal governments in addition to employers and those who make out-of-pocket health care expenditures. It follows that health care savings due to improved air quality are realized by all of these groups.

In what might be considered to be a modest proposal, one could conclude that even if the entire costs associated with complying with the mandates of the Clean Air Act were to be assumed by the federal government, the government would be ahead financially now and in the future.

References

1. US Environmental Protection Agency Office of Air and Radiation. The Benefits and Costs of the Clean Air Act:1990 to 2020. Washington DC: EPA, 2010.

2. US Environmental Protection Agency Office of Air and Radiation. Air Data: Generating Reports and Maps. Washington DC: EPA, 2011.

3. US Environmental Protection Agency. The Benefits and Costs of the Clean Air Act, 1970–1990. Washington DC: EPA, 1997.

4. US Environmental Protection Agency. National Ambient Air Quality Standards (NAAQS). Washington DC: EPA, 2011.

5. Office of Management and Budget OoIaRA. Informing Regulatory Decisions: 2003 Report to Congress on the Costs and Benefits of Federal Regulations and Unfunded Mandates on State, Locl, and Tribal Entities. Washington DC: Government Printing Office, 2003.

6. National Research Council. Hidden Costs of Energy: Unpriced Consequences of Energy Production and Use. Washington DC: National Academy of Sciences, 2009.

7. Levy JI, Baxter LK, Schwartz J. Uncertainty and variability in health-related damages from coal-fired power plants in the United States. Risk Anal 2009;29(7): 1000–14.

8. Epstein PR, Buonocore JJ, Eckerle K et al. Full cost accounting for the life cycle of coal. Ann NY Acad Sci 2011;1219:73–98.

9. US Environmental Protection Agency. eGRID, 2007. Available at http://www .epa.gov/cleanenergy/energy-resources/egrid/index.html. Accessed 2011.

10. Stern N. The Economics of Climate Change: The Stern Review. Cambridge: Cambridge University Press, 2007.

11. Council of Economic Advisors EOotP. The Economic Case for Health Care Reform. Washington DC: Government Printing Office, 2009.

12. Centers for Disease Control. Health Insurance Selected Characteristics: January—June 2010. Atlanta: CDC, 2010.

14
Policy Implications

"The Life You Save May Be Your Own"
—A short story by Flannery O'Connor

This is a book about health. More specifically, it is book about coal, its use, and how its use affects health. There are serious, adverse health effects associated with every aspect of using coal. Yet at the moment we depend on coal for almost half of the electricity used in the United States. Without this electricity, modern society could not thrive and the health care system would not exist. Our dilemma is that we need electricity, but we are clearly hurting ourselves now and in the future by using coal to generate it. Whether we are able to develop safe, clean, renewable sources of energy that pose minimal threats to health and the environment may be the most important determinant of the trajectory of our civilization.

Futurists, such as Arjun Makhijani, have envisioned a time when our energy needs will be met without nuclear or fossil fuels [1]. In his book he describes a path forward that is based on increases in the efficiency of energy use, maximization of solar energy production, and employing other technologies that are either available or foreseeable. These steps seem unlikely, given the current political environment. Too few people are aware of the health threats posed by coal-derived pollutants. Meanwhile powerful well-funded lobbies seek to eliminate or blunt regulations designed to improve air quality. All of this is complicated by a pervasive lack of knowledge about science and the way the physical world works, especially with regard to climate science, and a denial of the clear threats of global warming.

Nevertheless, there are causes for optimism. The quality of our air has improved as shown in figure 3.1 and further improvements are likely.

Perhaps most important, many ordinary people are making their voices heard in communities across the country.

Rulemaking—Agency Actions

Rules promulgated by federal agencies, including the Environmental Protection Agency (EPA), typically follow a standard series of steps as prescribed by the Administrative Procedure Act (5 USC §551, et seq.). This process is designed to ensure that stakeholders, including members of the public, are well informed and to give the public an opportunity to comment on the proposed rule. Steps in this process are designed to be open and transparent, thereby allowing the government agency, and possibly the courts, to evaluate the process and ensure that proper procedures have been followed.

The process begins when Congress passes a law. Typically the law either creates an agency or identifies the responsible existing agency and outlines the general goals of the law. The agency then creates rules that are designed to implement the law.

For many agencies, the next step, which is optional, is publication of an advance notice of proposed rulemaking. This step is designed to outline the rationale for a forthcoming rule and to set forth the initial analysis of the issues involved. The proposed rule is then published in the Federal Register, including the actual language of the proposed rule, the rationale that underlies the specific aspects of the rule, instructions for public comment, and a schedule of public hearings. The EPA typically schedules three public hearings at different locations across the country. Public comment periods typically last between 30 and 180 days. After modifications are made in response to comments, a final rule is posted, including a timeline for implementation. If the comments warrant substantial revisions to the proposed rule, a second, modified version may be posted followed by an additional period for public comment before promulgation of the final rule.

The process is designed specifically to make all steps open and transparent, and to take advantage of expertise brought to the process by the public. Of note, the "public" includes anyone not part of the agency involved. Individuals, organizations, and businesses are all parts of the public.

Rulemaking—Legal Actions

As is so characteristic of many aspects of modern life, the legal system is often used as a means to achieve an end. Environmental regulation is no exception. It is common for rules to be challenged in the courts.

In 2005, the EPA promulgated the Clean Air Interstate Rule. This rule was based on the Agency's finding that sulfur dioxide and oxides of nitrogen emanating from 28 states and the District of Columbia contributed to the failure to meet National Ambient Air Quality Standards for fine particles and ozone in downwind states. As discussed in chapter 3, burning coal is a major source of these two pollutants. The rule featured a market-based approach that created an optional cap-and-trade program for these two pollutants. A number of environmental organizations, states, and other entities challenged the rule alleging that it was not sufficiently protective and violated provisions of the Clean Air Act, forming the basis for the lawsuit that ensued. Some of these objections were technical in their nature. However, the petitioners complained that the rule failed to account for pollutants arising in each state and that it interfered with the acid rain program of the EPA, designed to reduce sulfur dioxide and nitrogen oxide emissions. The US Court of Appeals for the District of Columbia Circuit ruled in favor of the petitioners and found that the rule had numerous "fatal flaws." As a result the Court vacated the rule in its entirety. In other words, the challengers won. The Court directed the EPA to revise the rule in accord with its findings (No. 05–1244, DC Cir. 2008). The EPA proposed a response to the Court in July, 2010 in which it sought comments among alternatives designed to meet state-based emission limits [2]. The EPA estimated that in 2014 the rule would save between 14,000 and 36,000 lives, avoid 23,000 heart attacks, 26,000 emergency room visits, and a number of other benefits to health. Additional data relevant to the regulatory process were released in January 2011. The result of the process was the Cross-State Air Pollution Rule which was finalized on July 6, 2011.

Similar legal action was taken against the EPA concerning the Clean Air Mercury Rule. New Jersey was joined by other states and environmental groups in a suit that challenged the rule. Again, the court vacated the rule and required the EPA to develop a rule that was more protective of health. The EPA's response was posted in March 2011 [3]. The revised rule is designed to reduce air emissions of mercury and other heavy metals, such as arsenic and chromium. It would affect approximately 1,250 units that burn coal as well as others burning oil and natural gas. The goal is to reduce mercury emissions by 91% and to be implemented over four years. The EPA's website lists November 16, 2011, as the date for posting of the final rule.

On April 14, 2011, the EPA published a notice that outlined an agreement to settle a lawsuit against the Tennessee Valley Authority (TVA), a company that is, ironically, owned by the US government. This settlement is designed to require eleven TVA coal plants to modernize emission controls at an estimated cost of $3 to $5 billion, preventing between 1,200 and 3,000 deaths, 2,000 heart attacks, and 21,000 asthma attacks per year and thus saving $27 billion.

Many agree that the EPA rulemaking process "got it right" with its September 2010 final rule that regulates the Portland cement manufacturing industry (40 CFR parts 60 and 63). This rule is expected to yield $7 to $19 in public health benefits for every dollar spent by industry. Cement kilns are the third leading source of mercury emissions due to human activity. The new rule is expected to reduce mercury and particulate matter emissions by 92% and to reduce sulfur dioxide emissions by 78%.

The Clean Air Act and Air Quality Standards

In 1970 President Richard Nixon signed the executive order that created the EPA. This marked the beginning of the modern era in which the federal government began to play a seminal and active role in improving the quality of the air we breathe. Some of the details and a brief description of the history of the Clean Air Act are recounted in chapter 3.

As required by law, the EPA has conducted major assessments of the costs and benefits of the Act. The impact is shown in figure 14.1. The figure depicts air emissions and those that would likely to have occurred had the Act and the 1990 amendments not been enacted. The figure spans the times of an initial retrospective review of the Act and two prospective analyses. According to the second prospective review of the Act, the initial provisions of the act reduced air emissions to approximately half of what they would have been without the regulations imposed by the

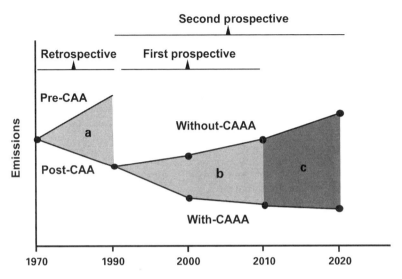

Figure 14.1
Impacts of the 1970 Clean Air Act (CAA) and the 1990 Clean Air Act Amendments (CAAA) on emissions. (a) As based on the retrospective analysis published by the EPA in 1997. A major reduction in emissions was predicted. (b, c) Projected emissions with and without the CAAA. The plateau in emissions occurring at about 2010 is the result of expected improvements in emission controls balanced by the effects of population growth. Note the nonlinear nature of the x-axis. Reproduced from the EPA Report: The Benefits and Costs of the Clean Air Act from 1990 to 2020 [13].

Act. Periodic revisions of air quality standards are required, as mandated by the Clean Air Act.

In its most recent report the EPA noted substantial improvements in air quality [4]. Compared to the 1990 baseline, the 8 hour 2007 levels were lower for the following pollutants:

- Ozone concentration, lower by 9%
- Annual fine particle (2.5 μm diameter) concentration, since 2000, lower by 11%
- Larger particle (10 μm diameter) concentration, lower by 28%
- Nitrogen dioxide concentration lower by 35%
- Sulfur dioxide concentration, lower by 54%

Reductions in the ozone and fine particle concentrations are expected to lead to the greatest benefits in health in the 1990 to 2020 interval, as shown in table 13.4.

Despite these improvements, the EPA estimated that over 125 million Americans lived in counties where one or more of its National Ambient Air Quality Standards were not met in 2008, as shown in figure 14.2 [4]. The failure to meet fine particle and ozone standards accounted for the majority of those individuals. Los Angeles bears the dubious distinction of having the most days when air quality reached the "Unhealthy for Sensitive Groups" criterion in the years 2001 to 2007 with an average of 100 days per year. Miami, Florida, Portland, Oregon, Minneapolis, Minnesota, and Seattle, Washington, had the best air quality in that interval averaging 3, 5, 5, and 6 days per year, respectively, when air quality was "Unhealthy for Sensitive Groups." As one might expect, the counties where air quality standards do not meet EPA standards areas are clustered around many major metropolitan areas, and include most of central and southern California, cities around the southern end of Lake Michigan and Lake Erie, cities along the Ohio River, and on the Atlantic Ocean between Washington, DC, and Boston.

In order to protect human health in an adequate manner, the Clean Air Act mandates periodic reviews of the National Ambient Air Quality

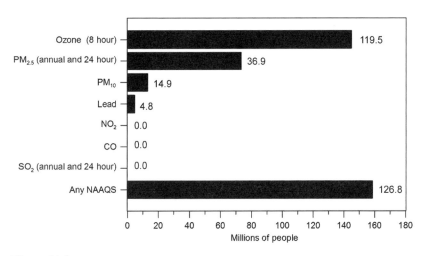

Figure 14.2
Number of people living in counties where pollutant levels exceed National Ambient Air Quality Standards in 2008. See table 3.1 for definitions of the standards. Reproduced from National Air Quality: Status and Trends through 2007 [4].

Standards. These are scheduled to occur every five years. The review process is supposed to permit the Agency to take advantage of advances in science and medicine in order to maximize benefits to the public. However, the reviews often do not take place as scheduled. As one might imagine, in a political system where the EPA is a part of the Executive Branch of the government, the procedures for air quality standard reviews may be politically driven. In May 2009, EPA Administrator Lisa Jackson issued a memorandum revising the process by which air quality standards were to be modified in order to make them less subject to political pressures [5]. In the most critical step she replaced the document that had been known as Advance Notice of Proposed Rulemaking (ANPR), a document written largely by EPA appointees, with a staff assessment. Administrator Jackson made this change because, in her judgment, the ANPR made the review process "vulnerable to the introduction of policy options that are not supported by the relevant scientific information." To help ensure access to the best scientific advice, the Clean Air Act created the Clean Air Scientific Advisory Committee. This committee and many environmental and public interest groups objected to the ANPR process on the grounds that it had become too political. The Staff Assessment was designed to restore scientific integrity to the review process while maintaining options for public participation and transparency in the air quality standards review process.

The controversy surrounding the revision of the air quality standard for ozone exemplifies some of these conflicts, described in part in chapter 8. At the time of the last revision in 2008, EPA Administrator Stephen Johnson set the 8 hour standard at 0.075 ppm (parts per million), ignoring the unanimous advice of the Clean Air Scientific Committee. The committee recommended a lower level. In the present revision of the proposed rule, the Agency indicated its intent to set the standard between 0.6 and 0.7 ppm. As a probable indication of the controversies surrounding this proposal, the agency has delayed promulgation of the final rule. A final announcement was anticipated on or before July 21, 2011. However, in September 2011, President Obama ordered the agency to withhold the final rule, a decision that pleased congressional Republicans and angered the health, scientific, and environmental communities.

The Role of Physicians

Why should physicians be especially concerned about coal and health? In a 1990 essay on physician involvement in environmental issues, McCally and Cassel argued that our special knowledge and expertise related to many elements of global environmental change conveys a responsibility to act on behalf of those affected [6]. These options range from personal actions and membership in and support for scientific and educational organizations, to encouraging and conducting research into the medical consequences of environmental pollutants. They argue that the fact that participation in the political processes may be required should not preclude action.

In an essay on medical professionalism, other authors defined three components: devotion to medical service, public profession of values, and negotiation or engagement with the public to establish social priorities [7]. The desire of doctors to engage in socially relevant activities as an element of professionalism was confirmed in a survey that showed that 90% of the responding physicians rated community participation, political involvement, and collective advocacy as being very important [8]. In that survey, 42.7% of those responding rated reduction of air pollution as being very important. In modeling physician responsibility and its relation to influences on health, Gruen et al. define domains of professional obligation and professional aspiration that include improving global health influences [8].

Although there have been advances in our ability to treat diseases known or suspected to have environmental links, prevention is always preferable to treatment. Primary prevention of pollutant-related diseases may require measures called for by the Precautionary Principle as put forth in the 1998 Wingspread Statement: "Where an activity raises threats of harm to the environment or human health, precautionary measures should be taken even if some cause and effect relationships are not fully established scientifically" [9]. A failure to act risks causing great harm if the threat is not dealt with. In the case of the health threats posed by pollutants derived from coal, randomized clinical trials are not possible, even though they are the "gold standard" in evidence-based medicine. Weaker evidence, in the form of epidemiological studies must serve as the basis for making decisions. The risks to health from allowing uncontrolled emissions are just too great and precautionary measures are required. This principle is echoed to a degree in the "First do no harm" aphorism derived from the Hippocratic Oath. The Precautionary Principle places the burden of proof on the putative polluter rather than physicians and our patients. It requires us to take action on pollutants that are highly likely to be harmful, unless or until the polluter can prove that the pollutant is harmless.

In a 1920 speech an early public health advocate had this to say about this emerging discipline [10]. "Public health is the science and the art of preventing disease, prolonging life, and promoting physical health and efficiency through organized community efforts for the sanitation of the environment, . . . which will ensure to every individual in the community a standard of living adequate for the maintenance of health." Much of this statement is still valid, even though our definition of health has been refined by the World Health Organization (WHO) [11]. This pioneer identified the following experts who were necessary for public health: physicians, nurses, social workers, epidemiologists, bacteriologists, statisticians, and engineers. As the requirements to meet the WHO definition of health expands, so must the list of experts. Climatologists, oceanographers, meteorologists, economists, and others now play important roles. The list is almost certain to expand with time and a growing understanding of the determinants of health.

Over the years many of the most important advances in health have been made possible by disease prevention as opposed to disease treatment. Prevention spares individuals from the rigors of disease and is almost always less expensive than treatment.

Policy Objectives

To paraphrase the Bible, "If coal offends, stop using it." This may not be easy. Nevertheless, due to the efforts of many who are concerned about the health impact of using coal, many coal plants are closing or being forced to modernize in the United States even as others are being built in China and other developing nations.

The amount of the pollutants emitted by modern coal-fired power plants is far lower than the amount emitted by the oldest plants. Even though carbon dioxide emissions continue unabated, the quality of the air is improving in the United States [4]. As ambient air quality standards evolve and become more protective, it is important to remember that "safe levels" for criteria pollutants and mercury do not exist. Every reduction in their concentration has demonstrable benefits to health and, as a bonus, saves money by reducing health care costs. Therefore it is imperative to educate the public and those who make decisions on the local, state, and federal level about the public health consequences of building new coal-fired plants and allowing existing units to avoid modernization. The message must be, "Coal-fired plants make people sick and die, particularly children and those with chronic diseases, and they cost society huge amounts of money desperately needed for other purposes."

In order to accomplish these objectives and force continued reductions in the pollutants produced by coal plants, it is important to be familiar with local, state, and federal regulations and procedures. These include rules and laws that govern the various phases of coal use, ranging from extraction through combustion and management of waste. On the local level, it is particularly important to be aware of zoning and permitting processes that govern the construction of coal plants, particularly portions of the permits that pertain to the proposed emissions of pollutants.

Controlling carbon dioxide emissions will be an extraordinarily difficult task. It pits the interests of individuals and societies in the long run against the immediate interests of governments and industry. The former need controls in order to maintain and improve current levels of health and the quality, and in some cases the continued existence of a sustainable environment, while the latter argue that economic growth requires access to the (falsely) cheap electricity produced by burning coal. The first step in solving this problem involves generating the will to confront the problem. The second step involves developing and deploying the technologies that will be required.

Because the climatological impact of a greenhouse gas persists for decades or even centuries after an emission, any policy that sets emission goals must look farther into the future than virtually any other policy in existence. Today's humans are not accustomed to thinking as far into the future as is required by our greenhouse gas problem. Pharaohs built pyramids to last forever, but too many corporate CEOs can only think ahead to the next stockholders' meeting and too many politicians do not think or see beyond the next election or campaign contribution. This must change.

The Intergovernmental Panel on Climate Change has developed several emission control scenarios [12]. The most aggressive of their scenarios, so-called Category I projection, estimates that stabilizing the carbon dioxide concentration between 350 and 400 ppm would require a reduction of emissions by 50% to 85% by 2050, relative to year 2000 emissions. Even if this were accomplished, temperatures would continue to rise between 2.0 and 2.4 degrees centigrade and ocean levels would rise by 0.4 and 1.4 meters due to thermal expansion of oceanic water. Larger increases of up to several meters could follow due to partial loss of the Greenland ice cap. Stabilizing the carbon dioxide concentration at 450 ppm, approximately 50 ppm above the current level, would require reductions in carbon dioxide emissions of around 25%, compared to 2000 levels, and is predicted to result in a global temperature increase of about 3 degrees centigrade and a rise in sea level of somewhere around one meter.

As a part of controlling pollutants, it is critically important to support the efforts of the EPA in its mission to protect human health and the environment. Powerful political forces, including some members of Congress and candidates for the presidency, seek to strip the EPA of its authority to regulate the emission of carbon dioxide and other air pollutants. Congressional actions absolving carbon dioxide as a cause of climate change will not change the fact that climate change is under way, human activity is the cause, and that carbon dioxide emissions are a major culprit.

To win the battle for clean air and prevent global warming it is necessary to adopt a message frame that emphasizes health and to confront the global warming deniers. If you take your sick child to a doctor and, after a comprehensive evaluation and series of tests, are told, "I'm terribly sorry, but your child has leukemia. However, with prompt treatment, we have an excellent chance for a cure." you might seek a second opinion. What if you went to 99 board-certified leukemia experts and were given the same advice, after a careful review and perhaps repetition of critical tests and then when you went to the 100th doctor and were told "I don't believe in leukemia; it is a hoax." what would you do? Global warming deniers believe the hundredth doctor. The results are predictable.

A global solution to the impending climate change catastrophe will not be possible without broad, international efforts. Developed nations have a special obligation to lead this effort and assist developing nations in their quest for economic growth while maintaining a safe, healthy, and sustainable environment.

It remains to be seen whether technology can "save" us. It won't happen by accident. It will take a sustained, determined effort to maximize the efficiency with which we use energy, to develop sustainable carbon-neutral sources of energy that do not pose threats to health and the environment, and, perhaps above all, to support the educational programs and scientific research that will make it possible to achieve these goals.

I am optimistic that we will succeed. The controls mandated by the Clean Air Act have prevented tens of thousands of deaths, even more serious illnesses, and saved billions of dollars. The atmospheric carbon dioxide record lends additional support to this optimism. Look again at the Keeling Curve, shown in figure 12.1. Each year the concentration of carbon dioxide dips as the earth tries to heal itself. This shows us that the earth still has the capacity to remove substantial amounts of this greenhouse gas from the atmosphere. Sadly, as the earth's healing processes come to their seasonal end, the carbon dioxide concentration rises to an even higher level. We can and must do better.

Data assembled by the Lawrence Livermore National Laboratory demonstrate that we waste about 58% of the energy we use each year in the United States and that 54% is wasted worldwide (Available at https://flowcharts.llnl.gov). Although the issues are massive and complex, it is important to avoid a state of paralysis that might be the result of a problem that seems to be beyond our grasp. Most experts agree that improving energy efficiency at every level is the least expensive way to move to a more sustainable energy future. On an individual level, improving home insulation, and using energy-efficient appliances, vehicles, and lightbulbs all make a contribution. Incentives to utilize solar, wind, geothermal, wave, and tide energy will create badly needed jobs, help improve the economy, and develop an energy portfolio that permits growth and health. Individual acts and the acts of organizations, companies, and political parties are all needed if we are to thrive as a society.

Final Thoughts

During my career as a neuroscience researcher and physician I have been privileged to observe and to investigate the intricacies of the human brain and how it functions. There is nothing like it.

In my medical school microbiology class we grew bacteria in Petri dishes. When a few microscopic bacteria are placed on the culture medium designed to foster their growth, they thrive and reproduce. The invisible single bacteria become colonies that are easily seen with the naked eye. As they grow, they consume nutrients and produce metabolic waste, but this is an unsustainable scenario. The continued survival of the bacteria becomes impossible as they exhaust their food supply and succumb to the toxins in the closed system of their environment. There is an analogy to be made with humans on planet Earth. Individual humans are invisible from space. However, lights from cities can be seen. We, like bacteria, live in a closed system. Bacteria don't have brains. They are not capable of choice. They are incapable of escaping their predetermined fate. As a society, will our brains save us? Are we smarter than bacteria?

The task of finding a path to a sustainable energy future, one that does not threaten our health and perhaps the survival of our species, will be long and difficult. But long and difficult does not mean impossible. Winston Churchill had this advice, "Never, never, never give up."

References

1. Makhijani A. Carbon Free and Nuclear Free: A Roadmap for U.S. Energy Policy. Takoma Park MD: Institute for Energy and Environmental Research, 2007.

2. US Environmental Protection Agency. Air Transport: Regulatory Actions. Available at http://www.epa.gov/airtransport/actions.html#jul10. Accessed 2011.

3. US Environmental Protection Agency. Clean Air Mercury Rule. Available at http://www.epa.gov/CAMR/. Accessed 2011.

4. US Environmental Protection Agency. Our Nation's Air: Status and Trends through 2008. Washington DC: EPA, 2010.

5. US Environmental Protection Agency. NAAQS Review Process. Washington DC: EPA, 2009.

6. McCally M, Cassel CK. Medical responsibility and global environmental change. Ann Intern Med 1990;113(6):467–73.

7. Wynia MK, Latham SR, Kao AC, Berg JW, Emanuel LL. Medical professionalism in society. N Engl J Med 1999;341(21):1612–6.

8. Gruen RL, Campbell EG, Blumenthal D. Public roles of US physicians: community participation, political involvement, and collective advocacy. JAMA 2006;296(20):2467–75.

9. Wingspread Conference Participants. Wingspread Statement on the Precautionary Principle. Available at http://www.gdrc.org/u-gov/precaution-3.html. Accessed 1998.

10. Winslow C-EA. The untilled fields of public health. Science 1920;51(1306): 23–33.

11. World Health Organization. World Health Organization Definition of Health. Geneva: WHO, 1948.

12. IPCC Core Writing Team. IPCC, 2007: Climate Change 2007: Synthesis Report. Contribution of Working Groups I, II and III to the Fourth Assessment Report of the Intergovernmental Panel on Climate Change. Geneva: IPCC, 2007.

13. US Environmental Protection Agency Office of Air and Radiation. The Benefits and Costs of the Clean Air Act: 1990 to 2020. Washington DC: EPA, 2010.

Glossary

Aerodynamic diameter: The diameter of a particle that has the same inertial properties (e.g., the terminal settling velocity, or the velocity at which the forces of gravity and drag are balanced) in air. For particles less than 0.5 μm (microns), the following equation can be used to calculate the aerodynamic diameter:

aerodynamic diameter = diameter of particle with same density and settling velocity(particle density)$^{0.5}$

Black lung disease: See coal workers' pneumoconiosis

Blood-brain barrier (BBB): Separates the circulatory system and the molecules, ions, and other elements in the blood from the brain, formed by tight junctions between the endothelial cells that line blood vessels and endothelial cell membranes. The BBB separates neurons from these constituents that would interfere with normal brain function.

British thermal unit (Btu): Approximately the amount of energy needed to heat one pound of water by 1 degree Fahrenheit

Carbon dioxide capture and storage (CO_2 CCS): A process consisting of the separation of CO_2 from industrial and energy-related sources, transport to a storage location, and long-term isolation from the atmosphere (from IPCC Special Report on CCS)

Cardiovascular disease (CVD): Any one of a variety of diseases affecting the heart or blood vessels or both

Chronic obstructive pulmonary disease (COPD): An irreversible progressive condition affecting the lungs that has features of emphysema and chronic bronchitis

Coal workers' pneumoconiosis: A chronic lung disease due to the inhalation of dust particles created by mining coal, also known as black lung disease

Confidence interval (CI): An estimated range that is likely to contain the variable of interest, usually set at 95%, thus if the 95% CI is 1.00–2.00, 95% of the time the variable would lie within that range. Small CIs are an indicator of a small amount of variance in the sample set.

Coronary heart disease (aka coronary artery disease): A condition in which plaque builds up in the coronary arteries, which are the arteries of the heart

Criteria pollutants: The 1970 amendments to the Clean Air Act required EPA to set National Ambient Air Quality Standards for certain pollutants known to be hazardous to human health. EPA has identified and set standards to protect human health and welfare for six pollutants: ozone, carbon monoxide, total suspended particulates, sulfur dioxide, lead, and nitrogen oxide. The term "criteria pollutants" derives from the requirement that EPA must describe the characteristics and potential health and welfare effects of these pollutants. It is on the basis of these criteria that standards are set or revised.

Disability-adjusted life years (DALY):   The DALY combines morbidity due to ill health and disability with early death into a single metric, expressed in years. It is used widely as a measure of disease burden by organizations such as the World Health Organization.

Enhanced oil recovery: The process where CO_2 is injected into existing oil wells in order to liberate additional crude oil

Evidence-based medicine: Conscientious, explicit, and judicious use of current best evidence in making decisions about the care of individual patients, and thus integrating individual clinical expertise with the best available external clinical evidence from systematic research

Forced expiratory volume in one second ($FEV_{1.0}$):   The maximum amount of air that can be exhaled, after a deep breath, in one second, a common measure of lung function and lung health

Hazard quotient: The ratio of the potential exposure to the substance and the level at which no adverse effects are possible. For an HQ greater than one, a risk is possible but not certain—can't be translated to a risk probability. For an HQ less than one, no risk is expected.

Heat content: The amount of heat energy available for release by complete combustion of a specified amount of coal, typically in millions of Btu per short ton (2,000 lb)

Implanted cardioverter defibrillator: A medical device permanently implanted in patients that senses potentially fatal cardiac rhythms, such as ventricular fibrillation, and delivers a shock to the heart muscle that is designed to restore a normal heart rhythm and prevent death.

Integrated gasification combined cycle (IGCC):   A multistage process wherein pulverized coal is injected along with steam to form CO and hydrogen. In a second step, additional steam is injected to yield CO_2 and hydrogen. Gas separation creates separate CO_2 and hydrogen gas streams.

Interquartile range: A measure of dispersion of individuals in a population, the difference between the first and the third quartiles.

Mountain-top removal valley-fill: A coal mining method in which the tops of mountains that cover a coal seam are removed and pushed into an adjacent valley; see chapter 4

Maximum contaminant level (MCL):   The legal threshold limit on the amount of a substance that is allowed in public water systems under the Safe Drinking Water Act

Odds ratio: The odds ratio compares the probability of an event between two groups. An odds ratio that is greater than one implies that the probability of an event is greater in the second group, usually the affected group. It is often presented along with a confidence interval that provides information about the amount of variance in the groups.

Progressive massive fibrosis: A severe form of CWP (coal workers' pneumoconiosis)

Pneumoconiosis: A chronic disease of the lungs caused by the inhalation of particles or fibers, such as coal workers' pneumoconiosis or black lung disease

Reference dose (RfD):   A numerical estimate of a daily oral exposure to the human population, including sensitive subgroups such as children, that is not likely to cause harmful effects during a lifetime. RfDs are generally used for health effects that are thought to have a threshold or low dose limit for producing effects.

Relative risk (RR):   The ratio of the probability of developing a condition between the exposed versus the nonexposed population

Total symptom scale (TSS):   A questionnaire developed to identify symptoms experienced during breathing experiments

Index